AUDITING A QUALITY SYSTEM FOR THE DEFENSE INDUSTRY

Charles B. Robinson

AUDITING A QUALITY SYSTEM FOR THE DEFENSE INDUSTRY

Charles B. Robinson

ASQC Quality Press
Milwaukee

AUDITING A QUALITY SYSTEM
FOR THE DEFENSE INDUSTRY
Charles B. Robinson

Library of Congress Cataloging-in-Publication Data

Robinson, Charles B
  Auditing a quality system for the defense industry / Charles B. Robinson.
    p. cm.
  Includes bibliographical references.
  ISBM 0-87389-078-7
  1. Munitions — United States — Quality Control — Auditing.
I. Title.
HD9743.U6R63 1990
623.4'068'5'20—dc20

ISBN 0-87389-078-7

10987654321

Acquisitions Editor: Jeanine L. Lau
Production Editor: Tammy Griffin
Cover design by Artistic License. Set in Century Schoolbook by DanTon Typographers.
Printed and bound by BookCrafters.

Quality Press, American Society for Quality Control
310 West Wisconsin Avenue, Milwaukee, Wisconsin 53203

Printed in the United States of America

TABLE OF CONTENTS

Preface ix

Introduction xi

Chapter 1   Defense Specifications/Standards Defining Quality Systems . . 1

       1.1 Defense Specifications/Standards 1
       1.2 Quality Program Requirements 2
       1.3 Quality System Requirements 2
       1.4 Basic Inspection Requirements 3
       1.5 Application of Quality Requirements 3

Chapter 2   Know the Specifications Before You Audit 5

       2.1 MIL-Q-9858A — Quality Program Requirements 5
       2.2 MIL-I-45208A — Inspection System Requirements 8
       2.3 MIL-Q-9858A and MIL-I-45208A Compared 10
       2.4 First — Know the Requirements 20
       2.5 Keep a Focus on the Basics 20

Chapter 3   Reason and Purpose of a Quality System Audit 23

       3.1 Procedures 27
       3.2 Compliance 28
       3.3 Assurance 28

Chapter 4   Audit Policy and Protocol 31

       4.1 Audit Policy 31
       4.2 Audit Schedule 32
       4.3 Supplier Audit 32
       4.4 Internal Audit 43

Chapter 5   Audit Worksheet 45

       5.1 General Information 45
       5.2 Audit Worksheets 47
       5.3 Worksheet Addendum 47
       5.4 Using the Worksheet 48
       5.5 Scoring the Audit 48

       5.5.1 The Evaluation Score 48
       5.5.2 Validate the Worksheet 49
       5.5.3 Evaluation Forms 49
       5.5.4 Has the Supplier Improved? 49
       5.5.5 Adequacy and Compliance Scoring 50
     5.6 How Is the Auditor Doing? 57

Chapter 6   Audit Reporting 59

     6.1 Who Prepares the Audit Report? 59
     6.2 Report Content 59
     6.3 Positive Findings 60
     6.4 Negative Findings 61
     6.5 Report Distribution 61
     6.6 Timely Reports 62
     6.7 Response to Audit Report 63

Chapter 7   Corrective Action and Follow-Up Requirements 65

     7.1 When to Request Formal Corrective Action 65
     7.2 Corrective Action Plan 65
     7.3 Follow-Up Action 66
     7.4 Close-Out Audit 67

Chapter 8   Auditor Qualifications 69

     8.1 Independence 70
     8.2 Personal Traits 71
     8.3 Education 73
     8.4 Experience 74
     8.5 Education/Experience Guideline 75
     8.6 Putting It All Together 76

Chapter 9   Product Audits 77

     9.1 Audit Scope 77
     9.2 Audit Frequency 77
     9.3 Quality Rating 78

Chapter 10 Special Process Audits 81

    10.1 Similarity to Quality System Audits 81
    10.2 Audit Policy and Protocol 82
    10.3 Worksheets 82
    10.4 Auditor Qualifications 83

Chapter 11 Corporate Audits 85

    11.1 Single Audits 85
    11.2 Information Exchange 85
    11.3 Needs and Strengths 86
    11.4 Overview of Corporate Audits 86
    11.5 Unified Supplier Base Audits 88

Chapter 12 Accept/Reject Criteria for a Quality System Audit Program . .89

    12.1 The Evaluator(s) 89
        12.1.1   The Customer 89
    12.2 Accept/Reject Criteria 90

Chapter 13 Terms and Definitions 93

Appendix A
    MIL-Q-9858A/AQAP-1 105

Appendix B
    MIL-I-45208A/AQAP-4 131

References 153

Index 155

PREFACE

The purpose of this book is to provide user-friendly information, procedures, and worksheets to those manufacturing companies supporting the defense industry. The audit of a quality system is gaining prominence in the evaluation of a company both within industry and by the government. The Contract Operation Review (COR) and the System Status Review (SSR) conducted by the Air Force and Defense Contract Administration Service (DCAS) have highlighted at least four significant needs that this book is designed to satisfy: auditor qualification, audit procedure, audit worksheets, and audit evaluation.

Auditor Qualification

The government's review of the defense industry indicates that either the auditor didn't identify significant areas of concern or management didn't take the auditor's report seriously.

Throughout this book, specifically in Chapter 8, the behavioral systems and technical requirements of an auditor are identified. We need to understand that auditors provide significant information upon which management decisions are made, and therefore we should employ meaningful qualifications for auditors. As the quality of auditors improves, management will more readily listen to their input.

The American Society of Quality Control (ASQC) has also recognized the importance of qualified auditors. Its Quality Audit Technical Committee, of which I am a member, has developed a Certified Quality Auditor program. An ASQC Certified Quality Auditor has the same professional status as does the Certified Quality Engineer and Reliability Engineer.

Audit Procedure

The government's review of the defense industry indicates that many of us had a less than adequate auditing procedure: We would be awakened by a problem, perform an audit, identify problem areas, and then go back to sleep.

There is more involved in an auditing program than just auditing. You need to know the requirements, why you are auditing, the policy and protocol, how to report audit results, when to ask for corrective action and follow-up, and how to assess the value of the audit as described in Chapters 2, 3, 4, 6, 7, and 12, respectively. There is much more to auditing than asking an auditor to complete a worksheet. I trust this book will effectively communicate the areas that need to be proceduralized to maintain a meaningful audit program.

Audit Worksheets

The government's review of the defense industry has effectively shown us the value of a worksheet that is created with a specific focus and used consistently.

The creation, use, and evaluation of a worksheet is explained in Chapter 5. The worksheet is a critical tool used by an auditor to ensure that the required areas are evaluated. This tool is necessary for thoroughness and consistency.

Audit Evaluation

The government's review of the defense industry, in my opinion, has failed to provide a meaningful evaluation. Red, yellow, and green lights or high, medium, and low significance are not specific enough to prioritize a corrective action plan. In some cases the government's audit report has taken months to issue and when the contractor responded it was as if the response went into the black hole.

Scoring, reporting, follow-up and closing out an audit, Chapters 5, 6, and 7, respectively, are areas that should get more than lip service. They are equally important to the auditor and auditee for prioritizing management action.

INTRODUCTION

Most people are tuned in to radio station WIIFM — what's in it for me

Before you read this book it is important that your perception of *auditors* and *auditing* is positive. The following analogy is appropriate because both could have life or death effects — one to human life, the other to business life.

My employer schedules me for a medical examination once each year. The physician who performs the examination and those who assist are highly trained to evaluate my present physical condition. After their review of all aspects of the examination they not only know my present physical condition, but are able to project my future health.

I follow a specific procedure to prepare for the medical examination, and a specially designed worksheet is used by those conducting it. The worksheet the medical professionals use is a guide that they follow; however, equally important is their personal knowledge, training, and experience.

One week after the examination I am called in to the doctor's office to hear the results. During that time I am not only told my present condition, but also areas that I need to improve on and what may happen if I don't heed the advice.

Areas that I question I bring to the doctor's attention so that they will be evaluated more closely. An interesting phenomenon occurs each time I am subjected to a medical examination. The doctor is able to point out areas I had never thought of that need to be improved to maintain my health. Fortunately, I listen carefully and attempt to make the necessary corrections.

An auditor performing an audit also needs training, experience, and a worksheet to examine an organization's health and identify areas that need improvement. Those who really care about the continued health of their organization should welcome a professionally staffed and organized audit regardless of how they perceive themselves.

Auditors and audits are like doctors conducting medical examinations: They are professionals who evaluate your current condition systematically and regularly and identify areas that are exceptionally good and those that need improvement. They do this so you are able to make adjustments that will help ensure your health.

I trust that if you don't currently share my positive opinion of auditors and the auditing profession, you soon will.

Total Quality Management

Your attitude toward auditing is a self-fulfilling prophecy — you choose

This analogy is the one that I prefer; however, there is another supporting reason why this book could be important to your health. The Department of Defense (DoD) is aggressively pursuing implementation of an initiative called total quality management (TQM).

TQM is a DoD initiative for continuously improving its performance at every level, in every area of DoD responsibility. Improvement is directed at satisfying such broad goals as cost, quality, schedule, and mission need and suitability. TQM combines fundamental management techniques, existing improvement efforts, and specialized technical skills under a rigorous, disciplined structure focused on continuously improving all DoD processes. It demands commitment and discipline. It relies on people and involves everyone.

Auditing is one of the fundamental management techniques that identifies areas for improvement and verifies adherence to a rigorous disciplined structure. Those companies that have an auditable TQM program will soon have a competitive edge in contract awards.

The DoD will develop criteria for evaluating continuous process improvement implementation. Source selection strategies will consider continuous process improvement as one element of selection. With these mechanisms developed to objectively evaluate efforts, continuous process improvement will be an essential capability for the defense industry. TQM will be applicable to prime contractors, subcontractors, and suppliers at all levels.

CHAPTER 1

DEFENSE SPECIFICATIONS/ STANDARDS DEFINING QUALITY SYSTEMS

The defense industry in the United States has almost exclusively used military and industry specifications that were generated and controlled within its national borders. Competition and an industrial technology exchange program have forced some companies to place contracts outside U.S. borders.

On the surface this doesn't seem to be a serious problem. All you have to do is require that the international suppliers conform to U.S. specifications. To impose these requirements on a company is costly and, in most cases, not necessary.

NATO's Defense Support Division has published several Allied Quality Assurance Publications (AQAP) that closely parallel the U.S. standards, and, in several cases, exceed the requirements. The NATO standards, which are better known internationally, have therefore been imposed on many of the international suppliers.

Because of the increase in international business, this book uses both the U.S. and NATO standards as a base on which to identify the quality system requirements that are to be audited.

1.1 Defense Specifications/Standards

There are many more specifications that augment and, in some instances, replace the ones on which this book is based. The following documents were chosen because they are the most widely used in the defense industry.

- MIL-Q-9858A, Amendment 1, *Quality Program Requirements*
- MIL-I-45208A, Amendment 1, *Inspection System Requirements*
- MIL-STD-45662A, Notice 3, *Calibration Systems Requirements*
- FAR Subpart 46.2, *Federal Acquisition Regulation's Contract Quality Requirements*

- AQAP-1, Edition 3, *NATO Requirements for an Industrial Quality Control System*
- AQAP-4, Edition 4, *NATO Inspection System Requirements for Industry*
- AQAP-6, Edition 2, *NATO Measurement and Calibration System Requirements for Industry*
- AQAP-9, *NATO Basic Inspection Requirements for Industry*

1.2 Quality Program Requirements

These are the most sophisticated requirements identified in this book. The documents that define these requirements are MIL-Q-9858A, FAR Part 46.202-3, and AQAP-1. Generally, this is reserved for those companies that have design authority and a sophisticated manufacturing/processing operation.

The two main differences between MIL-Q-9858A and AQAP-1 are that (1) MIL-Q requires maintaining a cost of quality program, whereas, the AQAP doesn't even mention it, and (2) the AQAP, via a reference to AQAP-6, requires that the cumulative effect of the errors in each successive component of a calibration chain be considered for each standard or measuring equipment calibrated.

The most significant requirements that are imposed by these two documents are:

1. Development of manufacturing work instruction.
2. Controlling of subcontractors at all phases of quality relative to services and supplies they provide.

The quality of the product or service relies heavily on the integrity of a quality program. Because of this reliance, an audit should always be performed to verify that the organization is in compliance with these requirements.

1.3 Quality System Requirements

These are the most widely used requirements in the defense industry. The documents that define these requirements are MIL-I-45208A, FAR Part 46.202-3, and AQAP-4. Generally, these are used by a company whose customers rely on their system to assure the quality of the product or service. Many processes cannot be fully or economically verified at receiving inspection; therefore, the purchaser relies heavily on a well-disciplined quality system at the supplier's facility.

The most significant difference between MIL-I-45208A and AQAP-4 is the cumulative error consideration during calibration that is imposed by AQAP-4, via a reference to AQAP-6 (reference 1.2).

The quality of the product or service often relies on the integrity of the quality system, especially when processing (welding, heat treat, plating, NDT, etc.) is performed. An audit of this system is usually performed primarily to assure that the processes are controlled properly.

1.4   Basic Inspection Requirements

These are the minimum requirements that are generally accepted in the defense industry. The documents that define these requirements are FAR Part 46.202-2 and AQAP-9. Simply put, these documents require that a supplier maintain a reliable inspection system and recorded inspection results. The basic inspection requirements are usually imposed on suppliers whose product or service can be fully inspected at receiving inspection. Because the product or service can be fully inspected at receiving inspection an audit is not normally performed on an organization that has a basic inspection system.

1.5   Application of Quality Requirements

The Federal Acquisition Regulation (FAR) Subpart 46.2 provides a table that may be used as a guide in selecting the appropriate quality requirement. The classification of the item or service being procured is the basis for selecting the quality requirement. The classification is determined by its *technical description, complexity,* and the criticality of its *application.*

Excerpts from FAR Subpart 46.2 along with some interpretation follow and a quality requirements guide is shown in Table 1.1.

Technical Description

- Commercial — described in commercial catalogs, drawings, or industrial standards.

- Military-federal — described in government drawings and specifications.

- Off-the-shelf — produced and placed in stock by a contractor, or stocked by a distributor, before receiving orders or contracts for its sale. The item may be commercial or produced to military or federal specifications or descriptions.

Complexity

- Complex — items that have quality characteristics, not wholly visible in the end item, for which contractual conformance must be established progressively through precise measurements, tests, and controls applied during purchasing, manufacturing, performance, assembly, and functional operation either as an individual item or in conjunction with other items.

Technical Description	Complexity	Application	Quality Requirement
Commercial	Noncomplex	Noncritical common	Basic
Commercial	Noncomplex	Noncritical peculiar	Basic
Commercial	Noncomplex	Critical	System
Commercial	Complex	Noncritical common	Basic
Commercial	Complex	Noncritical peculiar	System
Commercial	Complex	Critical	Program
Mil-Fed	Noncomplex	Noncritical common	System
Mil-Fed	Noncomplex	Noncritical peculiar	System
Mil-Fed	Noncomplex	Critical	Program
Mil-Fed	Complex	Noncritical common	System
Mil-Fed	Complex	Noncritical peculiar	Program
Mil-Fed	Complex	Critical	Program
Off-the-shelf	All	Noncritical	Basic
Off-the-shelf	All	Critical	System

Table 1.1 Quality Requirements Guide

- Noncomplex — items that have quality characteristics for which simple measurement and test of the end item are sufficient to determine conformance to contract requirements.

Application

- Critical — items in which a failure could injure personnel or jeopardize a vital agency mission. A critical item may be either peculiar, meaning it has only one application, or common, meaning it has multiple applications.

- Noncritical — items with any other application. Noncritical items may also be either peculiar or common.

- Basic — see paragraph 1.4.

- System — see paragraph 1.3.

- Program — see paragraph 1.2.

CHAPTER 2

KNOW THE SPECIFICATIONS BEFORE YOU AUDIT

The defense industry has been imposing military and NATO specifications on contracts with the erroneous assumption that they are understood by themselves and the contractor. Unfortunately, even the handbooks to the military specifications have not provided a uniformly understood interpretation.

It is imperative that, before a company launches into a meaningful audit program, a common understanding of the applicable specifications is reached. This should be done first internally and then the company's position should be understood and agreed upon by the cognizant government agency.

MIL-Q-9858A and MIL-I-45208A have been selected out of the many specifications that could be contractually imposed and broken down into manageable parts. This type of breakdown should be used as an outline to identify the procedures that satisfy each of the elements. A company's interpretation of the specification will then be established by virtue of the supporting procedures. The company is then prepared to explain its compliance to the specifications and obtain concurrence from its cognizant government representative.

2.1 MIL-Q-9858A — Quality Program Requirements

MIL-Q-9858A was issued December 16, 1963. Amendment 1 was released August 7, 1981 and Amendment 2 was released March 25, 1985. These amendments had some impact, but the document has remained basically the same since 1963. To clarify this military specification, the government issued Handbook H50 on April 23, 1965 that has remained unchanged.

Currently, MIL-Q-9858A and Handbook H50 are being updated, but this could take years. It is important, therefore, that a company carefully read and evaluate the impact of these specifications in today's environment.

The following is a quick overview of the items that need to be addressed when this specification is imposed on a company.

Quality Program Management

- The program must have clear objectives for its procedures, processes, and products.

- The program must be effective and economical.

- The design of the program must be based on technical aspects of products.

- The program must address design, development, fabrication, processing, assembly, inspection, test, maintenance, packaging, shipping, storage, and site installation.

- All products and services, both inside and outside of a contractor's facility, must be controlled by the program.

- The authority and responsibility of those in charge of all aspects of quality must be defined clearly by the program.

- Planning must take place at the earliest practical phase of the program.

- Work instructions must be used at each control point. Instructions must be:
    - Appropriate to the circumstances.
    - Monitored for effectiveness.

- Records of quality control must be maintained as objective evidence of the program.
    - Records must be complete and accurate.
    - Records must be useful for monitoring quality and initiating management action.
    - Actual values of nonconformances are required.
    - Actual values within tolerance are useful but not always required.

Corrective Action

- Must be implemented to detect and promptly correct assignable conditions which affect quality adversely.

- Must include analysis of data and examination of product.

- Must include analysis of trends in processes.

- Must include improvements in corrections and the monitoring of such actions.

Cost of Quality Program

- Cost of prevention versus correction must be evaluated by the contractor.

- Means of collecting and evaluating such data are at the contractor's discretion.

Facilities and Standards

- Effective document control must be practiced to ensure changes and revisions are implemented at effectivity points.

- Design and specification review must be practiced.

- Measuring and testing equipment must be maintained and controlled to assure accuracy.
    - Required equipment must be available.
    - All measuring and test equipment used to assure that samples conform to technical requirements must be calibrated per MIL-STD-45662.

Control of Purchases

- A system must be established with clearly outlined requirements for the selection of suppliers.

- The contractor's system shall provide for the clear transmission of requirements to its supplier.

- The system must include the evaluation/inspection of procured items.

- An effective means of implementing corrective action shall be practiced.

Manufacturing Control

- The quality program shall control all materials used in manufacturing by:
    - Performing receiving inspection on goods received.
    - Performing verification testing of materials received.
    - Maintaining identification and segregation of materials awaiting testing.

Control of Fabrication and Processing

- The quality program must ensure that adequate (written) work instructions, necessary equipment, and any required special working environments are provided by the contractor.

- The quality program must ensure that controls are implemented by physical inspection or process monitoring.

- The program must include measures for final inspection of products. Written inspection procedures must be used which clearly define accept/reject criteria.

- The quality program must include procedures for handling and storing the product. Procedures must address packaging and must be designed to prevent damage and loss.

- Periodic inspection must be performed to evaluate effectiveness of these procedures.

- Nonconforming material must be addressed by the quality program. Measures must be outlined for identifying, segregating, and disposition of such material.

- Statistical quality control may be practiced in the quality program. Sampling plans for inspection such as that outlined in MIL-STD-105 may be used.

- The program must include a system for displaying inspection status of products. Methods such as stamps, tags, boxes, or route cards are suggested.

- Certain contracts may require government inspection. When required, the quality program shall provide for such inspection in addition to the contractor's inspection. All documents shall be retained and made available for review.

- The program shall provide for the proper administration of government property. As a minimum, procedures shall provide for:

  – Inspection of property upon receipt.
  – Proper storage of property.
  – Proper maintenance of property.
  – Proper identification of property.
  – Proper usage of property.

As you can see this is a complex quality program. Before you approach your government representative with your audit program, you need to determine in your own mind and be able to convince yourself what are the accept/reject criteria when evaluating a MIL-Q-9858A quality program.

A careful reading of Chapter 5 and a self-evaluation using the worksheet (Appendix A) will help you understand this program and the implications/responsibilities associated with its implementation.

2.2 MIL-I-45208A — Inspection System Requirements

MIL-I-45208A was issued December 16, 1963. Amendment 1 was released July 24, 1981, which updated the reference of MIL-C-45662 to MIL-STD-45662, a calibration system requirement. This military specification also has a companion, Handbook MIL-HDBK-52 which was released July 7, 1964.

As with the specification reviewed in 2.1, this document must also be read carefully and evaluated to understand its impact in today's environment.

The following is a quick overview of the items that need to be addressed when this specification is imposed on a company.

Inspection System Requirements

- The system is designed to assure compliance to contract requirements through verification inspections of product.

- A documented inspection system is required. The system must include:
  - Written instructions with clear accept/reject criteria.
  - Records which describe both the nature and number of observations made along with a description of units.
  - A prompt corrective action mechanism for all assignable conditions which could affect quality adversely.
  - The system must include an effective means of document control.

- The system is designed to assure compliance to contract.

  - The system must provide for the availability of required equipment.
  - All measuring and test equipment used to assure that supplies conform to technical requirements must be per MIL-STD-45662.

- When required by specification or contract, process control procedures shall be a part of the inspection system.

- The system must provide a means of displaying the inspection status of product. Stamps, tags, boxes, or route cards are suggested.

- The system shall provide for the proper administration of government property. As a minimum, procedures shall provide for:
  - Inspection of property upon receipt.
  - Proper storage of property.
  - Proper maintenance of property.
  - Proper identification of property.
  - Proper use of property.

- Nonconforming material shall be addressed in procedures which provide for material identification, segregation, and disposition.

- Statistical sampling may be used in inspection. Such procedures and inspection methods may be used within this system provided they assure quality in the product.

- Government inspection may be required on certain contracts. When required, the contractor's system shall provide for such inspection in addition to the contractor's inspection. All documents shall be retained per contract and made available for review.

- The system shall provide for receiving inspection upon receipt of all procured supplies and materials.

This system, although less complex than the MIL-Q-9858A program, should still be studied thoroughly so that you have a firm knowledge of what an acceptable versus an unacceptable system looks like. Again, a careful reading of Chapter 5 and a self-evaluation using the worksheet (Appendix B) will help you understand this system.

2.3   MIL-Q-9858A and MIL-I-45208A Compared

A large number of defense contractors, particularly those doing business directly with the government on a major system, have a quality program that meets the requirements of MIL-Q-9858A. Although this sophisticated system is contractually imposed on them, it is neither necessary nor cost effective to flow the requirement to their supplier base.

More often than not they impose MIL-I-45208A, especially if the supplier is performing processes, the integrity of which cannot be verified fully at receiving inspection.

Quality audits are performed both internally and externally at the supplier base. It is important to both the internal and external auditor to know the differences between the two requirements. A sound understanding of both will prevent the auditor from requiring too little or too much when preparing for or conducting an audit.

The following comparison is keyed to the paragraphs of MIL-Q-9858A (MIL-Q) with MIL-I-45208A (MIL-I).

Quality Requirements Comparison Chart
(MIL-Q-9858A/MIL-I-45208A)

MIL-Q-9858A	MIL-I-45208A	REMARKS
1.1 Applicability	1.2.1 Applicability	Both state that when MIL-Q/MIL-I is specified it applies to all supplies or services referenced.
1.2 Contractual intent	1.1 Scope	MIL-Q quality program is subject to review and approval/disapproval by the government representative.
		MIL-I establishes inspection and tests necessary to substantiate product quality conformance.

MIL-Q-9858A	MIL-I-45208A	REMARKS
1.3 Summary	3.1 Contractor responsibilities	MIL-Q requires all supplies and services to be controlled at all points to assure conformance to the contract. Provisions are required for prevention, detection, and timely corrective action of detected discrepancies. Authority and responsibility of design, production, test, and inspection personnel shall be stated clearly. Program shall facilitate determinations of effects of quality deficiencies and quality on price. Program shall include an effective control of government property and purchased products or services. MIL-I system shall assure all supplies and services have been inspected and tested to meet contract requirements. The system shall be documented and available for review by the government representative before and during production. All changes to the system shall be made in writing to the government representative. In both MIL-Q and MIL-I government reserves the option to formally approve or disapprove the quality system.
1.4 Relation to other contract requirements	1.2.2 Relation to other contract requirements	Both similar except that MIL-Q has an additional requirement which states: "The contractor's quality program shall be planned and used in a manner to support reliability effectively."
1.5 Relation to MIL-I-45208A	1.2.3 Options	MIL-Q contains requirements in excess of MIL-I. MIL-I allows the contractor the option to implement the requirements of

MIL-Q-9858A	MIL-I-45208A	REMARKS
		MIL-Q, in whole or in part, provided there is no increase in price to the government. This option permits one uniform system in the event the contractor is already complying with MIL-Q.
2.1 Applicable documents	2.1 Applicable documents	MIL-Q references MIL-I and MIL-I references MIL-Q; they both require compliance to MIL-STD-45662.
2.2 Amendments and revisions	2.2 Amendments and revisions	Both allow the contractor to follow an official amendment to this specification provided the government is notified and there is not an increase in cost.
2.3 Ordering government documents	2.3 Ordering government documents	Both state that specifications, standards, and drawings may be obtained from the procuring agency.
3.1 Organization	N/A	Sufficient authority given to quality and management's regular review of the quality program. MIL-I has no counterpart.
3.2 Initial quality planning	N/A	Identify and make timely provision for the special controls, processes, test equipment, fixtures, tooling, and skills required for assuring product quality. MIL-I has no counterpart.
3.3 Work instructions	3.2.1 Inspection and testing documentation	MIL-Q states that all work affecting quality shall be prescribed in clear and completed documented instructions that are compatible with acceptance criteria for workmanship. These instructions shall also be monitored.

MIL-Q-9858A	MIL-I-45208A	REMARKS
		MIL-I states that clear and complete instructions for inspection and testing are required. Instructions shall include criteria for approval and rejection of product.
3.4 Quality control records	3.2.2 Records	MIL-Q requires records of inspection, test, and work performance monitoring shall be maintained and used as a basis for management action.
		MIL-I requires records of inspection and test.
3.5 Corrective action	3.2.3 Corrective action	MIL-Q is applicable to both product and quality program deficiencies. Product data and trend analyses are used and corrective actions monitored.
		MIL-I is applicable to products and services only.
3.6 Cost related to quality	N/A	MIL-Q states that cost of both the prevention and correction of non-conformances shall be used as a management tool.
		MIL-I has no counterpart.
4.1 Drawings, documentation, and changes	3.2.4 Drawings and changes	MIL-Q requires a procedure that will assure the adequacy, completeness, and currentness of drawings and to control changes in design. The contractor shall assure that the effectivity point of changes are met and that obsolete drawings and change requirements are removed from all points of issue and use. The contractor shall also be responsible for the review of process instructions, production engineering instructions, industrial engineering instructions, and work in-

MIL-Q-9858A	MIL-I-45208A	REMARKS
		structions relating to a particular design to assure their adequacy, currentness, and completeness.
		MIL-I requires that the contractor assure that the latest applicable drawings, specifications, and instructions required by the contract, as well as authorized changes thereto, are used for fabrication, inspection, and testing.
4.2 Measuring and testing equipment	3.3 Measuring and test equipment	Both are essentially the same in that equipment such as gages, measuring, and testing devices must be maintained to assure supplies conform to technical requirements. Both require compliance with MIL-STD-45662A.
		MIL-Q requires that prime contractor shall ensure that subcontractors and vendor sources also control accuracy of measuring and testing equipment.
4.3 Production tooling used as a media of inspection	3.3 Measuring and test equipment	Both are essentially the same in that they require devices used for inspection to be proved for accuracy at established intervals. MIL-STD-45662A applies.
4.4 Use of contractor's inspection equipment	3.3 Measuring and test equipment	Both are essentially the same in that both require inspection equipment be made available to the government to verify product quality.
4.5 Advanced metrology requirements	N/A	MIL-Q requires that timely identification and report to the contracting officer of any precision measurement need exceeding the known state of the art.
		MIL-I has no counterpart.

MIL-Q-9858A	MIL-I-45208A	REMARKS
5.1 Responsibility	3.8 Qualified products 3.1 Contractor responsibilities	Both are essentially the same in that they require supplies and services from contractors and subcontractors to meet contract requirements. Also, both require that qualified product list (QPL) items be inspected or tested to verify quality requirements. MIL-Q additionally requires the contractor to use to the fullest extent the objective evidence of quality furnished by their suppliers in evaluating the suppliers' ability to assure product quality. Also, the contractor shall establish a procedure for selecting qualified suppliers, transmitting applicable design quality and technical requirements, evaluating adequacy of procured items, and for early information feedback and correction of nonconformances.
5.2 Purchasing data	3.11.2 Purchasing documents 3.11.3 Reference data	MIL-Q requires that the subcontractors control all phases of quality relative to contracted products or services. All applicable requirements are included in purchase orders (PO). The PO shall contain a complete description of requirements for manufacturing, inspecting, qualification, or approvals. Technical requirements shall also be included. The chemical and physical test results on raw materials shall be provided. Instructions for direct shipments from subcontractor to government activities must also be included in the PO. MIL-I requires a statement to be placed on the PO to the effect that government inspection is required and that a copy of this PO shall be given

MIL-Q-9858A	MIL-I-45208A	REMARKS
		the government representative so that appropriate planning for government inspection can be accomplished.
N/A	3.10 Inspection provisions	MIL-Q has no counterpart. MIL-I allows alternative inspection, procedures, and equipment if concurrence from the government is obtained.
N/A	3.13 Government evaluation	MIL-Q has no counterpart. MIL-I allows the government to verify the conformance of the product and quality system to the requirements.
6.1 Materials and materials control	3.12 Receiving inspection	MIL-Q requires both the suppliers, materials, and products to be inspected upon receipt to assure conformance to technical requirements. Suppliers shall be required to exercise equivalent control of the raw materials used in the production of the parts and items which they supply to the contractor. Received material, not yet accepted, shall be segregated from accepted material. MIL-I requires that purchased supplies shall be subjected to inspection after receipt, as necessary, to assure conformance to contract requirements. Nonconforming government source inspected items shall be identified to the government.
6.2 Production process and fabrication	2.1 Inspection and testing documentation	MIL-Q requires that all production operations are accomplished using controlled conditions such as documented work instructions, adequate production equipment, and any spe-

MIL-Q-9858A	MIL-I-45208A	REMARKS
	3.4 Process controls	cial working environment. Criteria for approval and rejection shall be provided for all inspection of product and monitoring of methods, equipment, and personnel. MIL-I deals basically with clear inspection and testing instructions with criteria for approval and rejection of the product included. Process control procedures shall be an integral part of the inspection system when such inspections are a part of the specification or the contract.
6.3 Completed item inspection and testing	3.1 Inspection of completed supplies	MIL-Q requires final inspection and testing of completed products so as to sufficiently simulate product end use and functioning. This frequently involves endurance and qualification testing. MIL-I system shall assure all supplies and services have been inspected and tested to meet contract requirements.
6.4 Handling, storage, and delivery	N/A	MIL-Q requires work and inspection instructions for handling, storage, preservation, packaging, and shipping to protect the product at all times. This activity must also be monitored. MIL-I has no counterpart.
6.5 Nonconforming material	3.7 Nonconforming material	MIL-Q and MIL-I are identical with the exception that MIL-Q also requires the contractor to make available cost data associated scrap and rework (see MIL-Q para 3.6).

MIL-Q-9858A	MIL-I-45208A	REMARKS
6.6 Statistical quality control and analysis	3.9 Sampling inspection	MIL-Q and MIL-I are identical in that the sampling plan stated in the contract must be followed and if alternate sampling plans are used they shall be approved by the government. MIL-Q authorizes the use of MIL-STD-105, MIL-STD-414 or Handbooks H106, 107, and 108 provided the quality level of the product is not jeopardized.
6.7 Indication of inspection status	3.5 Indication of inspection status	Both require a positive system for identifying inspection status but it must be distinctly different from government identification.
7.1 Government inspection at subcontractor or vendor facilities	3.11 Government inspection at subcontractor or vendor facilities	Both document the right of the government to inspect product at any location.
	3.11.1 Government inspection requirements	
	3.11.2 Purchasing documents	
	3.11.3 Referenced data	
7.2.1 Government furnished material	3.6 Government furnished material	Both require the contractor to have procedures that control/protect government property.

MIL-Q-9858A	MIL-I-45208A	REMARKS
7.2.2 Damaged government furnished material	3.6.1 Damaged government furnished material	Both require the contractor to report damaged government property to the government.
7.2.3 Bailed property	N/A	MIL-Q requires procedures for the storage, maintenance, and inspection of bailed government property. MIL-I has no counterpart.
8.1 Intended use	6.1 Intended use	MIL-Q is intended to be used for the procurement of complex supplies, components, equipments, and systems for which the requirements of MIL-I are inadequate to provide needed quality assurance. Work operations, manufacturing processes, as well as inspections and tests are controlled. It assures interface compatibility among these units of hardware when they collectively comprise major equipments, subsystems, and systems. MIL-I is applicable to the procurement of supplies and services specified by the military procurement agencies.
8.2 Exceptions	N/A	MIL-Q is not applicable to the types of supplies for which MIL-I applies. MIL-Q does not normally apply to personal services and research and development studies of a theoretical nature which do not require fabrication of articles. MIL-I has no counterpart.
8.3 Order data	6.2 Order data	Both require procurement documents to identify the applicable MIL-Q or MIL-I specification.

2.4   First — Know the Requirements

This chapter should have not only clarified MIL-Q-9858A and MIL-I-45208A but, more importantly, made you aware of how a detailed study of the requirements is necessary before planning an audit. Many people in the defense industry claim to understand these *motherhood* specifications and conduct an audit ill prepared.

The results of being ill prepared are overcontrol or undercontrol, both of which are costly: pay me now (overcontrol) or pay me later (undercontrol), but you will pay.

This concludes a detailed review of just two of many documents. It is paramount that you appreciate the importance of preparing to plan an audit.

Know the requirements that apply to the area to be audited

2.5   Keep a Focus on the Basics

Today we work with sophisticated products that are produced with state-of-the-art technology, and we are proud of it. Even our language has become sophisticated (confusing), so much so that I don't think everyone knows what we are trying to communicate. Table 2.1 is an example of how we have identified meaningful programs in such a way that many of us don't know what they mean.

When reviewing new requirements (military standards, etc.) you will likely get confused. It is important to understand the practical value of each program so that you will be able to audit the effectiveness of them.

For example, I reviewed a company's cost of quality program that was impressive on the surface. The company not only identified the prevention, detection, and internal and external costs but broke these costs down to each work station. The data generated more than met the requirements of MIL-STD-1520; unfortunately, no one used the data.

The report was generated for one purpose; to comply with a contract requirement. The executives that received the report promptly filed it in "file 13."

To audit effectively, you must know not only the written requirement, but how to implement it effectively. In this example, the problem was that the company didn't know how to use the cost of quality to improve its quality and efficiency.

Today's Terms	The Basics
Total quality management Continuous improvement Best practices	Manage the company
Command media Procedures Work instructions	Who does it? How is it done?
System status review (SSR) Contractor operation review (COR) System worthiness analysis team (SWAT) System evaluation Systems audit Quality assurance system analysis	Verify
Incentive based corrective action (IBCA) Preliminary review board (PRB) Material review board (MRB) Get specified product compliance (Get SPEC)	Discrepant parts
Root cause corrective action Corrective action teams Corrective action boards Preventive action Defect analysis Statistical process control Quality improvement program	Fix it and prevent a recurrence
Total quality program	Comply with the requirements
Cost of quality	Efficient business practices

Table 2.1 Definition of Programs

CHAPTER 3

REASON AND PURPOSE OF A QUALITY SYSTEM AUDIT

The bitterness of poor quality lasts longer than the sweetness of on-time delivery

The president of a company establishes the management philosophy to be followed. This philosophy is then translated into a company policy and the president requires that all employees adhere to it. In small shops this policy may, unfortunately, be *understood* and not documented. For companies that comply with MIL-Q-9858A/AQAP-1, a policy, similar to the one that follows, should be formally issued and understood by the whole company. The company quality policy should also be used by the auditor when planning an audit. Areas of responsibility are clearly identified, and therefore sections of the audit worksheet may be directed to the appropriate individuals. Companies that comply with MIL-I-45208A/AQAP-4 or FAR 46/AQAP-9 also need a policy tailored to their system.

Quality Policy

Product Quality, Reliability, and Safety

It is the policy of the XYZ Company to design, manufacture, and market products and services which are: compliant with customer requirements and intended use; compliant with applicable government specifications, standards, and regulations; and safe for use by our customers and the general public.

It is also recognized that continuous efforts toward improvement must be built into every quality program in the company. Programs that do not consistently provide for ways to improve products, services, and quality systems soon become obsolete in our rapidly changing marketplace. Meaningful improvement results must be achieved if we are to continue to develop and to maintain our company as a high-quality, high-technology leader.

Purpose

The purpose of this document is to establish the framework and to assign responsibilities to assure that company products and services conform to the previously stated requirements.

Scope

This document applies to all departments within the XYZ Company. These departments may be involved in providing products and/or services. It must be emphasized that this document applies to all departments within the company and not just those with quality, reliability, or safety in their title.

Definitions

Quality — The totality of features and characteristics of a product or service that bear on its ability to satisfy stated or implied needs now and in the future.[1]

Reliability — The ability of an item to perform a required function under stated conditions for a stated period of time.[2]

Safety — The precautions taken during design, manufacture, inspection, testing, and the use cycle to preclude injury to personnel or property resulting from a discrepancy or malfunction in a product while it is operating in its contractual environment.

Responsibilities

The president of the XYZ Company is collectively responsible for the quality, reliability, and safety of all company products and services.

The director of engineering and the director of manufacturing operations, respectively, shall be responsible for designing and producing products that meet customer quality, reliability, and safety requirements.

The director of quality assurance shall coordinate a comprehensive quality program for all division functions, from design through delivery, at all manufacturing and overhaul locations and shall possess sufficient authority and organizational freedom to accomplish this task.

Engineering Design

Engineering design which provides for reliability, safety, and ease of transition into production is the cornerstone of product quality. Formal reliability studies should be made to ensure these objectives are met before release of new parts/products or release of major changes to existing parts/products. Manufacturing, quality, reliability, safety, and material should be part of the design review process for these parts/products.

Quality Assurance Functions

The quality assurance department shall have a functional and administrative relationship to management equal to that of other major departments. These relationships are mandatory to provide adequate organizational freedom for the accomplishment of quality objectives. Functions of the quality assurance departments include the following:

- *Establishment and maintenance of an effective quality assurance system.* This system must satisfy company and customer requirements adequately. The medium of communication and implementation shall be by a well-organized, clear, and current quality assurance manual.

- *Control of purchased materials, components, and parts.* Purchased materials shall comply to the same quality standards applied to manufacturing. This requires reputable suppliers capable of consistently providing products to acceptable quality levels. The supplier's system should provide evidence of capable processes and controls. Effective communications and contact with suppliers are required to assure that problems are known and corrected. The effectiveness of the supplier's quality system and processes is to be evaluated at some practical frequency. A system of source/receiving inspection shall be established to maximize assurance that all parts going into stock meet design/specification requirements.

- *Control of manufacturing and quality work instructions.* All productive work affecting product quality shall have documented work instructions as required by contract. Quality assurance shall assure that a system provides for the review and monitoring of the release of all new and modified work instructions used for production, test, and inspection.

- *Control of fabrication and assembly (inspection and test).* A system shall be maintained to verify product conformance after fabrication and assembly using direct inspection or approved statistical methods. No product shall be shipped or put in stock without verification of conformance to requirements. (See also *control of nonconforming materials* in this section.)

- *Defect prevention.* A quality assurance system shall be established, working with other departments to prevent defects from all levels of production and procurement.

- *Quality audit.* An audit program shall be established to assure compliance with procedures and work instructions affecting product quality and quality costs at all operating levels.

- *Monitoring customer requirements.* A system shall be established to monitor compliance with, and flow down of, all special quality requirements. This will include review of new contracts and contract changes by a formalized system to assure no special requirements are missed. This information will be made available to all pertinent activities.

- *Control of nonconforming material.* A system shall be established to control proper disposition of nonconforming material and the achievement of necessary action to correct the immediate problem and prevent recurrence.

- *Quality reporting.* A system shall be established for the collection, analysis, and dissemination of data pertinent to measurement of product quality and quality costs. Company and quality management shall act on this information to improve operations.

- *Control of metrology equipment.* Gages, test devices, and other measurement equipment must be provided in proper quantities and precision levels to assure consistent product conformance. The process operation, control system, and supporting services are part of this requirement. A documented system of sustaining accuracy, calibration, replacement of worn characteristics, inventory status, and location is part of this program. Measurement equipment must correlate with accepted standards. Where it is not feasible to eliminate the use of personal gages, these gages shall be controlled by the calibration system recall cycle.

- *Control of testing.* Acceptance testing must be accomplished by personnel assigned to the quality organization (or under the surveillance of quality personnel if performed by engineering or manufacturing). Personnel performing direct or delegated quality functions shall have sufficient, well-defined responsibility, authority, and organizational freedom to make decisions on the acceptability of a product based on test and inspection results.

- *Customer interface.* The quality assurance department shall provide customer reports, conduct necessary meetings with appropriate agendas and presentation materials, and visit customer facilities for routine and/or special problem-solving sessions as required. It shall also analyze, report, and establish any required corrective action on customer returns and warranty material. The quality function shall interface with the customer on quality matters.

- *Record retention.* Quality assurance records shall be maintained on file for a length of time appropriate to the product unless otherwise specified in contractual documentation.

- *Product recall.* A procedure shall be established and followed with respect to potential recalls of nonconforming material where such recalls are caused by suspect material.

- *Sales proposal support.* Quality assurance shall provide information to sales and contracts pertinent to: acceptability of the proposal's quality requirements, quality assurance and human resource inspection requirements, and facilities and equipment requirements.

Sound management practice, identified in this quality policy, requires checks and balances to assure important elements of the company are being performed properly and in order. Checks and balances not only make good business sense, but are also a requirement on MIL-Q-9858A. Paragraph 3.1 states ". . . management regularly shall review the status and adequacy of the quality program."

The president should periodically evaluate the total management system as well as the quality system. Both can be evaluated objectively by conducting an independent audit. A rude awakening awaits any president who is content to "assume" that the quality system is being properly implemented within the company or at a supplier.

We must remember that an audit is not intended to be a book of solutions, pat answers, or magic formulas identifying what should be done. The audit serves only to help surface critical issues and important symptoms that must be addressed for an operation to be, or continue to be, successful.

Seven Steps to Stagnation

1. We've never done it that way.
2. We're not ready for that yet.
3. We're doing all right without it.
4. We tried it once and it didn't work.
5. It costs too much.
6. That's not our responsibility.
7. It won't work anyway.

3.1 Procedures

The first thing an auditor does in preparing for an audit is to review the requirements that govern the area to be audited. Some company and most departmental procedures are simply a translation of requirements, not the requirements themselves.

The auditor must evaluate procedures in view of contract, government, and corporate requirements. This is a horrendous task if it has not been documented previously. Once the first evaluation has been completed, however, you will have established a baseline. The succeeding evaluations will need to review only procedural revisions and new or revised requirements.

Upon completion of the procedure review you will have verified the company's adequacy or identified areas that need correction. This step must be done before evaluating the company's compliance because individuals are not directed by the contract etc., but by internal procedures that interpret the requirements for them.

3.2   Compliance

This is the check and balance section of an audit. In 3.1 we simply identified and verified the adequacy of the procedures people are required to follow.

A worksheet must now be prepared (Chapter 5) so that the auditor will systematically cover all aspects of the quality system. Armed with the worksheet the auditor may now proceed in conducting an audit (Chapter 5). The results of this audit will let the general manager or quality manager know how well their quality directions are being implemented. It will also identify areas that need improvement.

Some people believe that they are so familiar with the requirements and experienced in auditing that the use of a worksheet inhibits them. A person with this attitude probably isn't qualified to perform an audit. Just as they look for procedures and instructions that direct individuals in the performance of a job, an auditor should use procedures and instructions (checklists) in the performance of an audit.

3.3   Assurance

The motive for performing an audit should be to assure that the quality system supports the company's objective of efficiently producing a cost-effective product or service that meets the customers' quality expectations.

Another motive, however, is to assure that the government will not find it necessary to intervene in the management of the company. The defense industry by its definition is important to the defense of the nation, and therefore it cannot and should not tolerate a company that, through bad quality, jeopardizes the national defense.

A regularly scheduled audit serves as an early warning signal that allows a company to take the necessary action to correct a deficiency and preclude it from recurring. If the government finds deficiencies, they have many distasteful and costly ways of informing management.

The Defense Contract Administrative Service (DCAS) refers to DLAM 8200.1, Part 5, Paragraph 4.502, when a deficiency is identified. Depending on the severity there are five ways of informing management.

1. Method A — This is when the defect is minor in nature and on-the-spot corrective action (CA) can be taken. Documented CA is not required in these cases.

2. Method B — This is issued when the defect is other than minor in nature, or when on-the-spot CA as to cause cannot be taken. Documented CA is required in these cases.

3. Method C — This is when a contractor has serious quality problems and/or an inordinate number of Method Bs are issued concerning an ongoing problem. A letter is forwarded to the contractor's top management requesting immediate action of the observed deficiencies and their cause. Depending on the nature of the deficiencies the contractor could be prevented from shipping items until the CA has been accepted by the government.

4. Method D — This is when the contractor cannot or will not comply with contract requirements and CA cannot be effected directly with the contractor by other methods. Often the result of a Method D is that all government procurement quality assurance activity is discontinued, which stops all government shipments.

5. Method E — This is when a subcontractor to a government prime contractor cannot or will not comply with contract requirements and CA cannot be effected directly with the subcontractor by other methods. Often the prime contractor is directed by the government to take immediate CA with the subcontractor. This is usually done by the prime contractor getting directly involved with the management decisions/CA plans of the subcontractor.

As you can appreciate, it is advisable to avoid any of these actions. The cost of an effective audit program will more than pay for itself if you only consider it as protection from the above DCAS or other agency corrective measures.

CHAPTER 4

AUDIT POLICY AND PROTOCOL

When a quality program (Chapter 2) is imposed, the defense industry requires written procedures to assure that every operation that has an effect on the quality of the product or service is consistently performed.

MIL-Q-9858A, paragraph 3.1 states: "Management regularly shall review the status and adequacy of the quality program." An effective quality systems audit program is an efficient way of *reviewing* your program. Additionally, a natural outgrowth of an audit is identification of situations that could cause deficiencies, and defect prevention should be the first line of attack. Defect correction is costly and often too late.

It is important at this point to recognize that an audit is not the only way to monitor quality. MIL-Q-9858A, paragraph 5.1 states:

> . . . The selection of sources and the nature and extent of control exercised by the contractor shall be dependent upon the type of supplies, his supplier's demonstrated capability to perform, and the quality evidence made available. To assure an adequate and economical control of such material, the contractor shall utilize to the fullest extent objective evidence of quality furnished by his suppliers.

What this basically says is that if you are able to adequately assure a product's quality by source/receiving inspection, then an audit may not be necessary. For those times when it is necessary or prudent to conduct quality audits, a proceduralized system should be established.

This chapter will discuss suggested policies and protocols for internal and supplier quality assurance system audits. From these policies and protocols you will be able to customize your own company procedure.

4.1 Audit Policy

A quality systems audit is not restricted to the auditing of a quality organization but should encompass the quality program, which may include every department and discipline within an organization. With this in mind the authority for conducting audits must come from, and be supported by, the chief executive of an organization. Without this umbrella authority, organizational cooperation and the value of the quality audit are limited.[3]

Support from the highest level of management is essential for quality audits to achieve their maximum benefit. The fact that audits are required by contract or specification means little if the requirement is not supported by management.[3]

4.2 Audit Schedule

A basic audit schedule should be generated and distributed at the beginning of each year in order for those affected to better plan their work to accommodate the audit. The schedule should consider audit planning time, follow-up audits, and unplanned audits. To schedule every day is unrealistic because during the course of a year a "mini crisis" is sure to take up some of the audit department's time.

4.3 Supplier Audit

Many of the details surrounding the total audit process are explained in Chapter 1. We do need, however, a general policy/procedure for conducting audits for two reasons.

1. We need to know what to do and how to do it to plan our work.
2. The supplier being audited needs to know how we are going to perform an audit.

The following is a sample of a company procedure for supplier audits. Some of the items the supplier need not be made aware of, but paragraphs VI.E. and VI.F. should, as a minimum, be communicated in advance of an audit so that the supplier will be able to prepare for the audit.

Procedure 4 — Supplier Evaluation Procedure

I.   Purpose

To establish a consistent method of evaluating, documenting, and maintaining records on supplier quality assurance evaluations.

II. Scope

This procedure applies to the performance of all supplier quality assurance system evaluations performed.

III. Authority

Company Policy No. 12.3, Quality Assurance Requirements

IV. Forms (Figures 4.1–4.4)

Survey/Audit Notification Letter
SE-1234 — Quality Assurance Supplier Status Report
SE-2345 — Survey/Audit Exit Interview Report
SE-3456 — Survey/Audit Deficiency Report

V. Definitions

A. Survey — (Preaward) The on-site evaluation of a supplier's quality capabilities to perform under the terms of a proposed subcontract.

B. Audit — A planned, independent, and documented assessment to determine whether agreed upon requirements are being met.

C. Evaluation — A systematic examination of the capability of an organization, or part thereof, to meet given requirements.[2]

D. Follow-up audit — An audit in which the purpose and scope are limited to verifying that corrective action has been accomplished as scheduled, and determining that the action was effective in preventing recurrence.

VI. Procedure

A. Survey/Audit Files

   1. The records are to be maintained for five years and shall include as a minimum the record of the last on-site survey/audit, if applicable.

   2. Records may be shown to, but not copied by, a government representative. Any copies shall be authorized by the procurement quality assurance manager.

   3. Review of the records by other customers shall be approved by the procurement quality assurance manager in advance.

ATTN: QA MANAGER   DATE _____
_____   SUPPLIER CODE _____
_____
_____

As a result of mutual agreement between your company and our Quality
Assurance Representative, a survey/audit is scheduled at your facility as
specified below.

DATE OF AUDIT _____

ARRIVAL TIME _____

REPRESENTATIVE'S NAME _____

REPRESENTATIVE'S PHONE # _____

AUDIT TYPE: SYSTEM [   ]   PROCESSING [   ]

A survey/audit normally takes 4–8 hours and is conducted as follows:

A. A brief introductory meeting with management and/or QA manager.

B. A review of your Quality Document(s) with the QA manager.

C. An in-depth review of your facility to determine compliance with the
   applicable quality specification(s). The review should be conducted with
   the QA manager or authorized knowledgeable designee.

D. A brief exit interview with management and/or QA manager to
   indicate the preliminary results of the audit. Final formalized results
   will be forwarded in approximately two weeks.

If you have any further questions, please contact the person below.

   John Doe
   Supplier Survey Coordinator
   (602) 555-1212

Figure 4.1 Survey/Audit Notification Letter

FORM NO. SE-1234

SUPPLIER _____ SUPPLIER CODE _____

ADDRESS _____ DATE _____

CITY, STATE _____ ZIP _____ PHONE (_____) _____

SUPPLIER CONTACT PERSON _____ TITLE _____

I. AUDIT TYPE: SYSTEM _____ PROCESS _____

       METHOD: ON-SITE _____ HISTORICAL _____ OTHER _____

II. STATUS

ACTIVITY	APPROVED	DISAPPROVED
INITIAL SURVEY	_____	_____
PERIODIC AUDIT	_____	_____
FOLLOW-UP	_____	_____
SPECIAL	_____	_____
SPECIFICATION		
_____	_____	_____
_____	_____	_____
_____	_____	_____
_____	_____	_____

III. APPROVAL TYPE

   LEVEL OF QUALITY ORGANIZATION

      QUALITY PROGRAM   _____

      QUALITY SYSTEM   _____

      BASIC INSPECTION SYSTEM _____

IV. COMMENTS _____

                  _____

                  _____

EVALUATION PERFORMED BY _____

Figure 4.2 Quality Assurance Supplier Status Report

FORM NO. SE-2345

SUPPLIER _____   SUPPLIER CODE _____

CONTACT PERSON _____   TITLE _____

DATE _____

☐   NO DEFICIENCIES IDENTIFIED

☐   INITIAL SURVEY/ADD PROCESS SURVEY — Below is a listing of quality system deficiencies identified during your quality survey. Correction of these deficiencies is required prior to your facility being considered for approval of your system/process.

☐   PERIODIC EVALUATION — Below is a preliminary list of the quality system deficiencies identified during our quality audit. A more detailed explanation of the deficiencies will be forwarded to you via a survey/audit deficiency report(s).

ITEM	DEFICIENCY/REMARKS

SUPPLIER ACKNOWLEDGMENT _____   TITLE _____
(receipt of this report)

SURVEY/AUDIT CONDUCTED BY _____

Figure 4.3   Survey/Audit Exit Interview Report

FORM NO. SE-3456 (front)

SUPPLIER NAME _____   SUPPLIER CODE _____

SUPPLIER CONTACT _____   REPORT DATE _____

SURVEY/AUDIT DATE _____   C/A DUE DATE _____

AUDITOR _____

The following deficiency requires a Corrective Action (C/A) plan/statement to be submitted by your company. Please respond to items 1 through 6 on the reverse side of this form on or before the due date indicated above.

REQUIREMENT

DEFICIENCY

FOLLOW-UP RESULTS		
	ACCEPT	REJECT

AUDITOR _____   DATE _____

Figure 4.4 Survey/Audit Deficiency Report

FORM NO. SE-3456 (back)

1. ROOT CAUSE OF DEFICIENCY
2. ACTION TAKEN TO CORRECT SPECIFIC DEFICIENCY
3. ACTION TAKEN TO CORRECT AND PREVENT RECURRENCE OF ROOT CAUSE
4. ACTION TAKEN TO DETERMINE IF OTHER PRODUCT IS AFFECTED BY SAME OR SIMILAR DEFICIENCY
5. CORRECTIVE ACTION EFFECTIVITY DATE(S)
6. SUPPLIER REPRESENTATIVE(S) RESPONSIBLE FOR IMPLEMENTING CORRECTIVE ACTION

SUPPLIER SIGNATURE _____ DATE_____

DISPOSITION OF CORRECTIVE ACTION — (Customer Use Only)					
				ACCEPT	REJECT
FOLLOW-UP AUDIT REQ'D?	YES	NO	IF YES, DATE REQ'D		
SIGNATURE _____			DATE _____		

Figure 4.4 (cont.)

B. Evaluation Schedule

1. Every supplier shall be evaluated every eighth quarter, regardless of the level of its quality system. The requirement for an audit is as follows:

   1.1. Basic inspection requirements (Chapter 1) — a survey is not required.

   1.2. Quality system or program requirements (Chapter 1) — an initial survey is required. The need for an additional audit will be reviewed only if a significant quality problem is detected, the quality rating falls below an acceptable threshold for two consecutive months, or product has not been received for more than four quarters.

   1.3. Quality system or program requirements that also include Processes — an initial survey plus a survey to the specific process is required. Additional audits will be conducted once every eighth quarter, as a minimum, depending on their quality history. (See 1.2 for unacceptable quality history.) When a survey is conducted a representative process from each applicable process family (welding, heat treat, plating, nondestructive testing, etc.) will be evaluated.

C. Evaluation Criteria

1. The following criteria are used to consider the waiving of an audit.

   a. Products/Services can be fully inspected at receiving inspection.

   b. The last audit had an initial score of greater than 70%.

   c. The supplier has been inactive for over four quarters. In these cases you should consider dropping the supplier from your approved list.

2. The following criteria are used to consider performing an audit.

   a. The last audit had an initial score of less than 70%.

   b. The quality rating, determined by receiving and source inspection, has decreased below the acceptable threshold (95% acceptance rate) for the second month.

   c. A supplier audit is contractually required. These special requirements must be communicated to the procurement quality assurance organization by the contract review personnel.

3. Once these criteria (2a–c) have been evaluated, the procurement quality management must determine if an audit will be performed.

D. Evaluation — Without On-Site Activity

   1. The product/service can be fully inspected at receiving inspection.

   2. The supplier does not have significant quality assurance problems based on the continued receipt of products.

E. Evaluation — Audit of Existing Suppliers

   1. Preparation

      1.1. Call the supplier to arrange for a convenient time to conduct the audit. During the conversation inform the supplier that a worksheet will be sent that needs to be filled out before the audit.

         NOTE: The requirement for the supplier to fill out the worksheet may be waived if sufficient response time will not be available before performing the audit.

      1.2. Send the supplier the audit worksheet (Chapter 5) along with a Survey/Audit Notification Letter (Figure 4.1) identifying the time, estimated length, and number of auditors involved in the audit.

      1.3. Review the completed worksheet and the suppliers' history file (last audit, C/A responses, quality rating, etc.).

   2. In-Briefing

      2.1. Attendees: Quality assurance manager, manufacturing manager, and plant manager. At a minimum the quality assurance manager should be present.

      2.2. Explain the scope and mechanics of the audit.

      2.3. Set up a tentative time for the exit interview.

   3. Conduct the Audit

      3.1. The areas to be covered are defined in paragraphs B.1.1. through B.1.3.

      3.2. Assure that someone representing the supplier, preferably the quality assurance manager, accompanies you during the audit.

      3.3. Keep your supplier escort informed during the audit so that they will not be surprised during the exit interview.

3.4. Document deficiencies as they are detected. Much of the detail may be lost if you wait until the end of the day to document your findings.

4. Exit Interview

4.1. If deficiencies are found, document them on form SE-2345, verbally identify them during the exit interview, answer any questions, and then ask the supplier to acknowledge receipt of the exit interview report by signing it.

5. Corrective Action

5.1. The items noted on form SE-2345 (Figure 4.3) shall be evaluated as to the need for corrective action. Those needing corrective action should be documented on form SE-3456 (Figure 4.4) and forwarded to the supplier.

5.2. Deficiencies that indicate fraud or conditions that will definitely have an adverse affect on the quality of products/services shall be immediately communicated to the auditor's management.

5.3. Response by the supplier to the SE-3456 deficiency reports should be received by the auditing department by a specified date. The date is determined by adding 34 days to the date the reports were mailed (30 days plus four days mail time).

5.4. The need for a follow-up audit should be determined after the receipt of the supplier's response.

6. Audit Completion

6.1. The audit cannot be completed until all outstanding corrective actions have been resolved satisfactorily.

6.2. Form SE-1234 (Figure 4.2), Quality Assurance Supplier Status Report, should be generated and maintained in the record file.

F. Evaluation — Survey of New Suppliers

1. Preparation

NOTE: When warranted by operational considerations the requirements of paragraph 1.3 may be waived.

1.1. Identify what product/service the prospective supplier will perform and what quality standard the quality organization will be required to meet (Chapter 1).

1.2. Call the supplier to obtain permission to perform a survey. Explain that you will be sending a copy of the survey form to be completed before the audit.

1.3. Send a letter (Figure 4.1) to the supplier identifying what needs to be done in preparation for the survey.

    a. Complete the applicable sections of the survey form and return it to the surveying department.

    b. Send a copy of the current quality assurance manual along with the completed survey form.

1.4. Upon receipt of the completed worksheet and the quality assurance manual, determine if the supplier has an acceptable quality system. This evaluation may result in one of the following actions:

    a. Schedule a survey because the supplier appears to have a fully compliant quality system.

    b. For minor deficiencies, send a letter to the supplier requesting that the deficiencies be corrected before the survey.

    c. For more serious deficiencies, send a letter to the supplier requesting that they correct the deficiencies and document the actions taken for the survey teams review before scheduling the survey.

    d. For quality systems that appear to be far below the required standards, consider identifying a different supplier. If you absolutely need that particular supplier follow step 1.4b. If you are able to consider another supplier send the unacceptable supplier a letter informing them that their quality system is not compatible with your needs and, therefore, they will not be considered for addition to the approved supplier list.

2. In-Briefing (similar to E.2)

3. Conduct the Survey

3.1. The areas to be covered are identified in paragraph B.1.2 through B.1.3, except that processes shall be individually approved.

3.2. Assure that someone representing the supplier, preferably the quality assurance manager, accompanies you during the survey.

4. Exit Interview

4.1. If deficiencies are found, document them on form SE-2345, verbally identify them during the exit interview, answer any questions, and then ask the supplier to acknowledge receipt of the exit interview report by signing it.

4.2. Explain to the supplier that the corrective action responsibility is theirs and the evaluation process will not continue until after they respond in writing to each of the noted deficiencies. No response on their part will indicate they do not wish to do business with the company.

5. Corrective Action

5.1. If a corrective action response is received it will be evaluated for the need to conduct a follow-up audit.

6. Survey Completion

6.1. The survey cannot be completed until all outstanding corrective actions have been resolved satisfactorily.

6.2. Form SE-1234, Quality Assurance Supplier Status Report (Figure 4.2), should be generated and maintained in the record file.

4.4 Internal Audit

Some people treat strangers (suppliers) with more respect and courtesy than they do those within their own company. There are special considerations when going to a supplier's facility, but we should accord many of the same courtesies when performing an internal audit.

The sample Supplier Evaluation Procedure may be modified slightly and become an Internal Audit Procedure. As you may have noticed I changed the title from *evaluation* to *audit* as the evaluation does not necessarily require an audit. Internally you should not make allowances to *waive* an audit. You may need to reschedule one from time to time, but they should always be performed.

Although most internal quality assurance audits are performed by an organization within the quality assurance department, I believe it should be done by a separate organization that does not have a direct responsibility in the production schedule. In large organizations this could be part of the financial department or a separate management systems department that reports directly to the general manager. For small organizations I believe it should not be done by the quality manager, but by the general manager.

An audit schedule should be generated that assures that each area is evaluated on a yearly basis. This is important for all concerned. It allows both the auditor and the auditee to plan in advance. Audits typically cover the "cradle to the grave" starting with contract/purchase order negotiation and ending with after market evaluation. Some of the departments that are evaluated include: contracts, engineering, material, manufacturing engineering, manufacturing, purchasing, quality engineering, inspection, assembly, customer support, etc.

An Auditor's Protocol

An auditor is like a guest in a home and should treat the auditee with the same respect and courtesy offered a host

CHAPTER 5

AUDIT WORKSHEET

Two sample worksheets are given in this chapter: MIL-Q-9858A/AQAP-1 and MIL-I-45208A/AQAP-4. I had some reservation about including any worksheet because it is critical that you validate any worksheet before adopting it. The tendency is to "speed-read" or give a cursory review and then use it because it is contained within a book and has been successfully used as an auditor training tool within the defense industry.

You should avoid this tendency: those who don't will find out, too late, that it didn't totally meet their specific needs. The worksheets are essentially the same as those used by a major aerospace company, but I have modified them slightly to accommodate the defense industry in general.

Rather than bouncing back and forth between internal and external (supplier) survey/audits, the comments in this chapter are slanted toward external audits. As mentioned in Chapter 4, you need only make slight modifications to apply this internally because the principles and process are essentially the same.

5.1 General Information

A cover sheet asking general information questions should be attached to each worksheet. The supplier should complete this before the audit for two reasons: to save time and to give the auditor an understanding of the size and complexity of the facility to be audited.

Figure 5.1 is an example of the type of information to obtain. This will allow you to be better prepared when conducting an audit.

You will know:

- If they have a quality manual and if it is a dynamic or stagnant document. (The revision date will give you a hint.)

- How sophisticated they believe their system to be. Until your supplier base understands the thoroughness of your audit, many of them will overstate the level of their quality system.

1. ***ORGANIZATION***

   CONTACT _____ TITLE _____

   QA MANAGER _____ PHONE # _____

     TENURE OF QA MANAGER ___ YEARS

   TITLE OF QUALITY DOCUMENT _____

     CURRENT REVISION _____ REVISION DATE _____

   QUALITY SYSTEM: MIL-Q-9858A _____ AQAP-1 _____

                  MIL-I-45208A_____ AQAP-4 _____

                  FAR 46.201-1 _____ AQAP-9 _____

2. ***GENERAL DATA***

   TOTAL # OF EMPLOYEES _____ QA _____ MFG _____

   TOTAL PLANT AREA: _____ Sq. Ft.

   BUSINESS MIX: _____ % MILITARY _____ % COMMERCIAL

3. ***PRODUCT CAPABILITY***

   SUPPLIER'S PRIMARY PRODUCT OR SERVICE _____

   ADDITIONAL PRODUCT/SERVICE(S):

   Castings _____ , Forgings _____ , Electronics _____ , Gears _____ ,

   Machine/Grind _____ , Pressure Test _____ , Sheet Metal _____ ,

   Other _____

   _____

   SPECIAL PROCESSES PERFORMED:

   Brazing _____ , Welding _____ , Soldering _____ , Heat Treat _____ ,

   NDE (PT ____ MT ____ RT ____ Other ____), Plating/Coating ____ ,

   Calibration _____, Other _____

   _____

Figure 5.1 General Information

- How large an organization they have and the ratio of quality to manufacturing personnel. The average ratio ranges from 1:7 to 1:12 depending on the sophistication of your product.

- What type of disciplines will need to be audited. This will aid you in establishing the qualifications of the audit team/person.

5.2 Audit Worksheets

The determination of what quality requirement to use has been explained in Chapter 1. Sample worksheets for the most common ones are included in Appendix A (MIL-Q-9858A/AQAP-1) and Appendix B (MIL-I-45208A/AQAP-4).

These worksheets are training tools, intentionally written to cover each area as thoroughly as possible, and therefore they contain much redundancy; i.e., indication of inspection status is covered six times in MIL-Q and seven in MIL-I. As you become more proficient the worksheets can be streamlined.

At the beginning of each worksheet there are instructions for the supplier. They are asked to identify their particular procedure(s) that satisfies questions identified with a *Code 1*. This needs to be done in advance for two reasons:

1. It saves time during the audit because the supplier should know their procedures far better than the auditor. If this is not done in advance of the audit the auditor will have to spend up to four hours at the supplier's facility just reading and identifying applicable procedures. This is a waste of time both for the auditors and suppliers.

2. It allows the supplier to be better prepared for the audit and possibly to correct areas in advance.

I recently sent a MIL-I-45208A worksheet to a supplier to complete. I wanted to conduct an audit of their quality system as soon as possible because of some customer concerns. I called the supplier and asked how soon they could be ready and was told a couple of weeks as the worksheet pointed out some procedures they needed to create. The audit was successful because of their advance corrections.

5.3 Worksheet Addendum

The supplier's history should be reviewed to identify specific areas of concern that should be evaluated during the audit. Some of the resources you may want to query are receiving inspection history, the supplier's last audit and associated corrective action commitments, and the performance of the items supplied either in your plant or by the user.

Once you have identified your concerns, develop an addendum to the standard checklist. This specialized worksheet will increase the value of the audit by verifying that the supplier does in fact implement meaningful corrective actions.

5.4   Using the Worksheet

Some people believe that a worksheet is just a guide and it is inhibiting to follow step by step. They assume that an experienced auditor will ask all the essential questions without using a worksheet. To give the audit credibility it should be consistent, which demands that a worksheet be used.

People without much experience need to use it consistently otherwise they will get too involved evaluating a specific problem area and forget to cover other areas.

An experienced auditor still needs to use the worksheet, but may not need to follow it step by step. For myself, I go into a specific area and ask questions based on my experience and knowledge of the worksheet. Before I leave the area I review the worksheet, item by item, to assure I have covered every area and have documented the results.

Before I leave, but after I have verified that all questions have been covered, I discuss my observations with my escort and the person in charge of the area. This helps me to correct misconceptions on my part and/or prevents the supplier from being surprised during the exit interview.

5.5   Scoring the Audit

Scoring a survey or audit allows you to:

- Validate the worksheet so that it more closely correlates with the quality of hardware received.

- Determine if the supplier has improved since the last evaluation.

- Evaluate a supplier in relationship to others of similar size and complexity.

- Stay current with the activities of the auditors and suppliers.

5.5.1   The Evaluation Score

This is simply a percentage of the applicable questions that were answered acceptably. The evaluation of XYZ Machine Shop (Figures 5.2 and 5.3) indicates that their overall quality program rating is 75.5% or satisfactory,[4] but the rating of their calibration system is 61.1% or poor.

The supplier needs to respond to all unacceptable areas, but a special emphasis should be placed on calibration (61.1%), nonconforming material (66.7%), and corrective action (66.7%). The corrective action requirements will be more thoroughly discussed in Chapter 7.

5.5.2 Validate the Worksheet

The whole purpose of performing a survey or audit is to make a judgment of the suppliers probability of producing quality hardware on time. It stands to reason, therefore, that you would want the evaluation rating to have some correlation to the quality of hardware inspected at receiving inspection.

To validate a worksheet you first need to accumulate three to six months of completed worksheets. Then you will need to divide them into supplier categories: it is unfair to compare worksheets/receiving inspection results from a casting, electronic component, and packaging supplier. Each supplier category may have its own peculiar fingerprint which would cause you to modify the worksheet individually. Please don't confuse this with process surveys/audits which are addressed in Chapter 10; I am only addressing quality systems/programs.

If you are unable to adjust the worksheets to correlate with receiving inspection results you may need to look at the auditor qualifications (Chapter 8).

5.5.3 Evaluation Forms

Blank forms for evaluating a quality program (MIL-Q-9858A/AQAP-1) or quality system (MIL-I-45208A/AQAP-4) are included (Figures 5.4 through 5.7).

5.5.4 Has the Supplier Improved?

You need a quick and reliable method of determining the status of your supplier base quality systems. Using individual categories and total system percentage ratings makes this determination possible without requiring a person to scan a multipage checklist.

The bar chart (Figure 5.3) is a quick visual aid and the data given in Figure 5.2 are easily adapted to a computer for trending in an infinite number of ways. You can quickly determine if suppliers have improved from their last rating and how they compare to the average rating of suppliers in their particular category.

5.5.5   Adequacy and Compliance Scoring

The scoring method just reviewed is a broad overview of a supplier and is easily calculated. A more sophisticated scoring system would still use Figures 5.2 and 5.3, but two sets of charts would be generated: one for procedure adequacy and the other for procedure compliance.

Procedure Adequacy

An evaluation of the supplier's procedures is needed to verify that they have procedures that address the contractual requirements. Then an evaluation of those procedures is done to verify that they adequately satisfy the contractual requirements.

MIL-Q-9858A, paragraph 3.3, states that "The quality program shall assure that all work effecting quality shall be prescribed in clear and complete documented instructions of a type appropriate to the circumstances." The review and separate scoring of procedure adequacy allows you to verify this requirement, and also gives you assurance that the supplier properly understands the requirements and has made an effort to communicate them in a reliable and consistent format.

Procedure Compliance

A supplier may have a "perfect" set of procedures, but if they are not adhered to the whole quality system falls apart. The actual floor audit is where the auditor will spend the most time at the supplier, and it is there that a supplier's discipline (adherence to procedures) is made evident.

Using this double-scoring method identifies how well an organization is managed. If the procedures are not adequate, but they follow them, then you only need to have the procedures corrected; you have the confidence that the supplier will enforce and follow written procedures. However, if the procedures are good, but they do not follow them, you have a serious problem. That supplier doesn't need to change procedures, which is relatively simple; they need to change the whole culture/discipline, which is exceedingly slow and difficult.

I was on an audit team that was asked to evaluate a company that had significant problems. By significant problems I mean they had to fire the plant manager and most of his immediate staff for fraudulent reasons. Their procedures had been totally rewritten and the new quality manager was starting to have a positive effect, but it was slow. He told me that his biggest problem was trying to change the "OTC" (old tribal custom).

EVALUATION CATEGORIES	EVALUATION RATING				
	Total	N/A	Accept	Reject	% Accept
A. Quality Program	20	4	13	3	81.3
B. Drawings, Document, and Changes	16	1	12	3	80.0
C. Calibration	36	—	22	14	61.1
D. Sampling Inspection	3	—	3	—	100.0
E. Purchasing	19	2	13	4	76.4
F. Receiving Inspection	7	—	6	1	85.7
G. Stores	6	—	5	1	83.3
H. Fabricating/Processing	13	1	9	3	75.0
I. Final Inspection	7	—	6	1	85.7
J. Shipping	6	—	5	1	83.3
K. Nonconforming Material	3	—	2	1	66.7
L. Corrective Action	9	—	6	3	66.7
M. Cost of Quality	6	—	6	—	100.0
N. Government Property	8	4	3	1	75.0
TOTAL	159	12	111	36	75.5

Figure 5.2 XYZ Machine Shop — Evaluation Categories

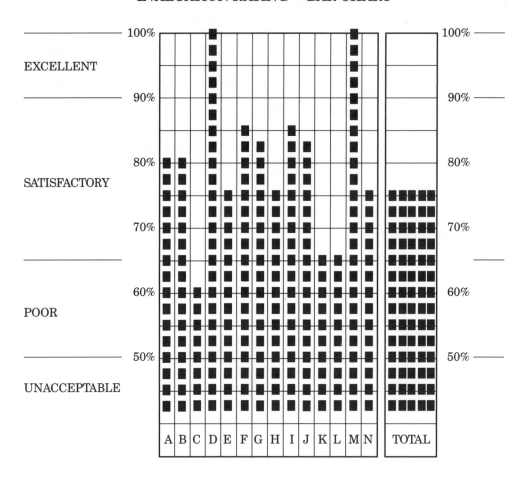

Figure 5.3 XYZ Machine Shop — Evaluation Rating

MIL-Q-9858A, Amendment 1 and AQAP-1, Edition 3

	Company Surveyed/Audited	Date

EVALUATION CATEGORIES	EVALUATION RATING				
	Total	N/A	Accept	Reject	% Accept
A. Quality Program	20				
B. Drawings, Document, and Changes	16				
C. Calibration	36				
D. Sampling Inspection	3				
E. Purchasing	19				
F. Receiving Inspection	7				
G. Stores	6				
H. Fabricating/Processing	13				
I. Final Inspection	7				
J. Shipping	6				
K. Nonconforming Material	3				
L. Corrective Action	9				
M. Cost of Quality	6				
N. Government Property	8				
TOTAL	159				

Figure 5.4 Quality Assurance Program Survey/Audit

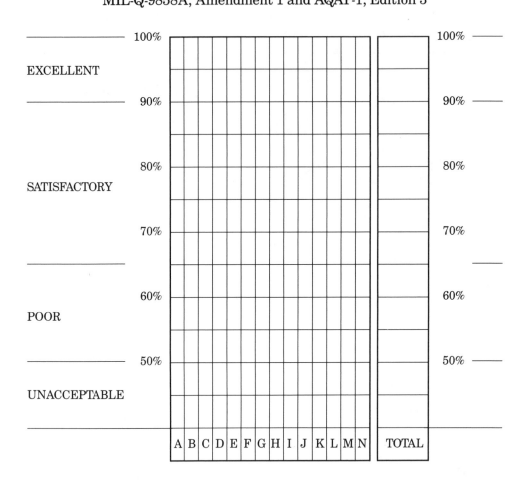

EVALUATION RATING — BAR CHART

MIL-Q-9858A, Amendment 1 and AQAP-1, Edition 3

Figure 5.5 Quality Assurance Program Survey/Audit — Evaluation Rating

MIL-I-45208A, Amendment 1 and AQAP-4, Edition 4

Company Surveyed/Audited				Date

EVALUATION CATEGORIES	EVALUATION RATING				
	Total	N/A	Accept	Reject	% Accept
A. Quality System	22				
B. Calibration	36				
C. Sampling Inspection	4				
D. Purchasing	7				
E. Receiving Inspection	14				
F. Stores	4				
G. Fabricating/Processing	10				
H. Final Inspection	4				
I. Shipping	5				
J. Nonconforming Material	4				
K. Corrective Action	8				
TOTAL	118				

Figure 5.6 Quality Assurance System Survey

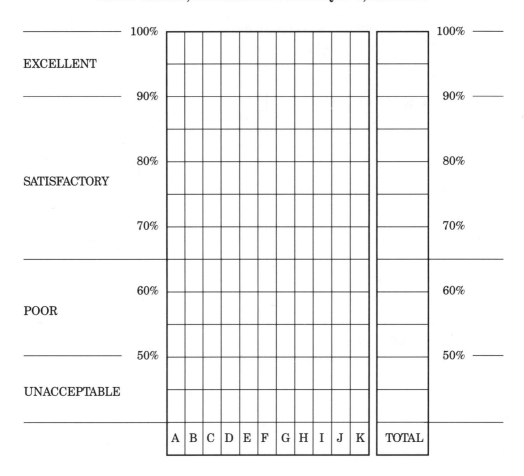

Figure 5.7 Quality Assurance System Survey — Evaluation Rating

5.6   How Is the Auditor Doing?

Unfortunately, this has a negative connotation, but it is meant to be of value not only to management but to the auditor. The worksheets demand that both the supplier and the auditor complete sections that are difficult, if not impossible, to be done in the confines of an office.

When management reviews the survey/audit worksheet they can generally determine if the auditor is proficient in evaluating a supplier's quality system. Try as we may we will sometimes ask people to do things they are not qualified to do. This insight will allow management to initiate meaningful training programs and/or shift a person's area of responsibility.

Over a period of time management knows about how long it will take to complete a particular evaluation at a supplier. The average time an auditor takes and its relationship to the other auditors may point up a need. If the time is considerably shorter, then the credibility of the results may be questioned, and if the time is much longer than average, the person may be too picky or not qualified.

CHAPTER 6

AUDIT REPORTING

The content of an audit report must be objective and its distribution restricted. Needless damage to an organization's reputation and/or morale and possibly lawsuits could result if this discipline is not maintained. However, if suppliers, purchasing, or department heads are not aware of negative findings they will not be able to effect the necessary preventive/corrective action. How to document the results and decide who should and who should not receive the audit report is a fine and sometimes sensitive line.

6.1 Who Prepares the Audit Report?

The lead auditor conducts the exit interview (Chapter 4) and is also responsible for preparing the audit report. The audit report is, in a sense, the exit interview formally documented.

Although it is ultimately the lead auditor's responsibility, before releasing the report the auditors involved in the particular audit should review the contents to verify its objectivity and accuracy.

I remember the embarrassment of issuing a formal audit report that I thought was accurate only to find out that a significant finding was not accurate. The auditor had referenced a finding number from a previous audit that had not been corrected. Unfortunately, I restated the audit finding from a different previous audit. Two mistakes were made: (1) the auditor didn't express the deficiency clearly, and (2) I didn't ask the auditor to review the final report before issuing it.

The purpose of the report is to communicate important information. It should be written so that people with reasonable knowledge are able to understand it without asking a series of questions.

6.2 Report Content

- Auditee's name — internal department or supplier name.
- Dates during which the audit was performed.

- Scope of the audit — quality program, quality system, basic inspection (Chapter 1), specific process, special, etc.

- Key personnel contacted during the audit — this allows the auditee to contact the person within the organization who knows the circumstances around the finding.

- Audit team members — identify the lead auditor.

- Checklist — identify the checklist/procedure/document used to guide performance of the audit.

- Results of the audit — both positive (6.3) and negative (6.4).

- Rating — a bar graph (Figure 5.3) should be included for system surveys/audits.

- Appreciation — thank the auditees for their cooperation.

- Response date — the day in which a corrective action response will be due and identification of who the response is to be sent to. Usually this is required within 15 to 30 days after the audit report date.

- Distribution — list of the individuals that will receive the report and their respective positions (6.5).

6.3   Positive Findings

All too often we concentrate on the negative and do not give credit where credit is due. The XYZ Machine Shop (Chapter 5) had a 100% rating in both sampling inspection and cost of quality categories. The audit report should document this degree of excellence. Additionally, if they had received a 90% or greater rating in another category, that too should be recognized.

I have read several books that recommend against specifying the actual survey/audit rating as it causes debate and dissension internally as well as with the customer and supplier. The establishment of a rating, in my opinion, is profitable for it allows everyone to monitor progress using the same criteria. If you cannot justify/support the rating, then perhaps the rating needs to be modified. The rating does give the auditees a means of measuring their ability to meet the requirements; a goal.

Those who do not have a goal will reach it every time

6.4 Negative Findings

The audit report should be a general overview of the negative findings with only the most important ones documented. Attached to the report, however, should be a copy of all the findings.

For the XYZ Machine Shop (Chapter 5) particular mention should be made of the calibration (61.7%), nonconforming material (66.7%), and corrective action (66.7%) categories.

A summary is important because not everyone may have the time to read and digest each of the individual findings; however, they still have a need to understand the results. Additionally, it makes for a quick reference when someone asks you a question about a particular department or supplier.

Each finding should identify what effect the continued practice will have on the quality of future products, activities, or services. This should be stated as objectively (don't embellish the projected effect on quality) as possible without being judgmental.

Proprietary Information

Negative findings that cannot be documented without divulging proprietary information must be documented separately. These items should be coordinated with the auditee to assure that positive corrective action will be initiated. Whether or not it is documented needs to be handled on a case-by-case basis.

Fraud, Waste, and Abuse

Findings that may indicate fraud, waste, or abuse should be referred immediately to your legal department. After a positive determination that supports this possibility, the incident must be documented and referred to the appropriate government agency. Failure to do this could result in criminal charges.

Items of this nature would obviously be documented in a separate letter and only included in the audit report with the concurrence of your legal staff.

6.5 Report Distribution

An audit report is a sensitive document and the distribution must be evaluated regularly to assure that only the necessary individuals are on distribution.

Distribution lists remind me of mail order lists; once your name has been added to the computer you will be on distribution forever, even if you ask to be removed.

Unfortunately, I am on some of those lists that are generated internally and by the societies to which I belong. An audit report is too sensitive to be treated in this manner.

The people on distribution must be able to have an impact or be impacted by the results of the audit; i.e., individuals that can contribute to the corrective action process. Table 6.1 identifies what positions should be considered for inclusion on the distribution list.

Supplier Audits	Internal Audits
Supplier Distribution	Department manager
	Department quality manager
President	Quality director
Quality manager	Manufacturing director
Manufacturing manager	Engineering director
Engineering manager	Internal audit manager
Internal Distribution	
Procurement QA manager	
Purchasing manager	
Buyer	
Supplier audit manager	

Table 6.1 Positions Included on Distribution List

6.6 Timely Reports

The government conducted a *Control of Subcontractor Audit* at one of our companies, but didn't promptly issue the formal audit report. It took them six months to issue it and during that time we repeatedly asked for information about the audit, but nobody was willing to talk to us. Accompanying their report was a requirement to respond within 10 working days.

If the report is not issued within two weeks the auditee has an indication of the importance you put on it. Additionally, the quality of the report is put in jeopardy as the audit team may not be able to accurately clarify questionable findings if too much time lapses between the audit and the report.

6.7 Response to Audit Report

A response to the audit report should be requested within approximately 30 days. To eliminate people playing games with due dates the report should specify a day (November 20, 1988) by which the response should be received. To calculate the day add estimated mailing times for you and the auditee to 30 days from the date the report is mailed. The auditee should be asked to include a response/corrective action plan as well as a time schedule for resolution of each item.

CHAPTER 7

CORRECTIVE ACTION AND FOLLOW-UP REQUIREMENTS

Corrective action and follow-up may not always be required. The severity of the deficiency coupled with the corrective action response needs to be considered.

7.1 When to Request Formal Corrective Action

To determine if a formal corrective action response is required you must look first at the deficiency's effect on the product or service being rendered and then at any special contract requirement. If the deficiency puts either the product/service or contract compliance in jeopardy then a formal corrective action needs to be obtained.

Deficiencies that may not meet these criteria but are systemic in nature should also be treated in the same manner. More than the actual deficiency, a systemic problem indicates a lack of control that could filter into more significant areas.

To ask for a formal corrective action plan for a grammatical error in a procedure or one document out of 100 being obsolete is a waste of time for both the auditor and the auditee. A request such as this also reduces the meaningfulness of the audit to a "witch hunt."

7.2 Corrective Action Plan

To develop a corrective action plan you must first identify the deficiency and then determine the root cause. The root cause of a dimension being machined out of tolerance is not always an *operator error.* In fact, it is probably caused by one or more of the following circumstances:

- Machine malfunction
- Improper heat treating
- Tool setting error
- Deficient work instructions

• Environmental conditions

Once the deficiency and root cause has been positively identified, the corrective action plan can be developed. The plan should include what was done to correct the specific deficiency and what action has been taken to preclude a similar occurrence on that and similar products. The plan should identify not only the action to be taken but who is responsible for that action and by what date the preventive action will be in effect.

A corrective action request comes all packaged up with a due date, which most of the time can be met. However, some deficiencies take longer to investigate. In these cases a response should be sent that identifies the progress you have made so far and the date by which you will submit your corrective action plan. Don't get caught in the trap of responding quickly without a thorough investigation; more often than not the government will see through a plan that is poorly conceived.

We must have the mind-set that a meaningful corrective action plan will benefit the auditee more than the auditor. Too many people see them as just another road-block that prevents them from being productive.

7.3   Follow-Up Action

A review of the corrective action response and its timeliness is, in some instances, all the follow-up action that is necessary. This review, however, may indicate that the auditee is not responsive or is evasive ("We have determined corrective action is not necessary"), or the action is inadequate ("The employees have been counselled to refrain from this practice in the future").

When the response is questionable or unacceptable you should immediately contact the auditee in an attempt to resolve your perception. A simple telephone call could resolve your concern by just clarifying the corrective action plan. Acceptance of the plan should be given if there is a reasonable chance that it will correct the problem and preclude a recurrence: your way is not the only way.

On serious issues where submitted documentation (revised procedures, work instructions, etc.) will not verify the corrective action effectiveness, a follow-up audit may be warranted. For suppliers that are not geographically convenient you may elect to send a field source inspector or a contract auditor to verify the implementation. If the deficiency warrants a follow-up audit, don't let the geographic location of supplier sway your decision to perform one.

7.4 Close-Out Audit

Upon resolution of each finding a written notification should be sent to those who were on distribution of the initial report. This not only satisfies the auditing department's records, but also lets the auditee objectively know that the corrective actions have been accepted.

There are occasions when the response has gone through many iterations without being resolved. When this deadlock is reached the audit should be closed out. This is not a resolution, but a notification to the auditee of what action has been taken as a result of the company's inability to resolve the problems satisfactorily.

This impasse may result in disapproval or increased involvement with the supplier to assure the quality of the product or service being purchased. Your response may be similar to a "Method E" (Chapter 3).

CHAPTER 8

AUDITOR QUALIFICATIONS

A few years ago an auditing function was a part-time job. It was usually hardware oriented so management asked an inspector to conduct audits. Today, we know that an auditor evaluates not just hardware, but also systems and procedures. The part-time auditing job has become an auditing profession. The auditor must know the discipline being audited, be able to develop a checklist, communicate in a non-threatening manner to the shop floor personnel, make judgments on a compilation of information, present formal exit briefings to senior management, document audit results in an objective manner, and many other subtle things that, if not done in a professional manner, could cause more harm than good.

A company may have a perfect audit policy, procedure, and checklist, but qualified persons performing audits are equally important. If a company spends all of its effort on developing a system without screening and training its audit staff adequately, it will have completed less than 50% of the needed preparation. A worksheet is a guide that needs to be followed, but a competent auditor should be able to do more than mechanically follow written guidelines.

Auditing requires both human relations and technical skills. Human relation skills are hard to define. As you read this chapter keep the following saying in the back of your mind and you will better understand "human relation skills."

That which offends the ear will not easily gain admission to the mind, and that which offends our ego will arouse nothing but anger and indignation

Technical skills in this context include general knowledge as well as expertise that is specific to the type of defense industry in which a person is involved: electronics, pneumatics, mechanical, distribution, etc. A person's resume is important, but the thorough personal interview is critical. Too often a thorough appraisal is not made before promoting individuals. This tendency has led one student of occupational patterns to formulate the "Peter Principle," which states that in a hierarchy every employee tends to rise to their level of incompetence, and "Peter's Corollary" which states that in time, all posts tend to be occupied by employees who are incompetent to carry out their duties.[5]

Selecting individuals to perform internal or external audits is hard and serious work. The people in these positions have an affect on a company's reputation and management decisions made both within a company and by customers and suppliers.

8.1 Independence

The auditor's independence rests on the absence of conflict of interest in organizational relationships, or in the mental attitude[6] of the auditor with respect to the audit and the possible ramifications the audit report may have on the company/department or personnel.

This is usually not a problem when auditing a supplier. By definition the auditor does not have any organizational relationship internally or externally. Most companies conducting business in the defense industry require their employees to sign a conflict of interest disclosure document that would prevent a person with any interest in a supplier to have any influence in the company's relationship with it.

Internally, most companies place the internal audit department within the quality assurance organization. The theory behind this is that this organization's charter is to verify objectively that quality products or services are being shipped to the customer. This organization is in compliance with government regulations, but too many companies hold quality responsible for discrepant hardware and late shipments. With this pressure the reality of the quality assurance department's independence and objectivity is somewhat questionable.

I believe that the optimum place for an internal audit department is in a separate organization that reports directly to the CEO. We espouse that a quality program involves all departments within an organization, not just the quality assurance department, yet we single out just one department to conduct audits. Having the audit department report directly to the CEO has many benefits:

1. Executive management has direct insight into the actual operation and contractual compliance of the company. It is not watered down through many levels of management.

2. The auditors have organizational independence that allows them to report objectively without preference to the quality organization or any other department.

3. The company knows objectively that the CEO is really interested in its compliance to company and contractual requirements. The direct reporting of the audit results to the CEO causes managers to be more attentive to their responsibilities.

4. The CEO has an objective evaluation of the effectiveness of each department. This allows resources to be placed in needed areas and/or managers to be promoted or demoted.

8.2 Personal Traits

How auditors present themselves is directly proportionate to the success of the audit. Not only the auditor's but the company's or department's credibility can be severely damaged if they do not conduct themselves in a professional manner.

Auditor selection should be made with the following attributes in mind:[3]

- *Ability to listen.* The department or supplier being audited knows more about their operation than you do. You will learn more listening than talking; an auditor should be a student not a teacher. A good listener not only patiently hears the words but is able to understand the thought or idea being communicated.

- *Ability to communicate.* Regardless of how intelligent an auditor is the auditee will not respect the input from a person who has bad grammar or from one who cannot communicate a thought clearly. This is not in reference to a slight speech impediment or a regional or ethnic accent; this is in reference to verbal and written skills that precisely communicate a thought or idea to a person or group of persons.

- *Ability to plan.* An auditor cannot afford to "wing it." Careful planning of the audit schedule and checklist, and preanalysis of the auditees current system and previous audit results are critical to a meaningful audit.

- *Ability to lead and control.* Auditors should not be soft-soaped by "nice guys," intimidated by "bad guys," or unduly impressed by "big meetings" where executives try to impress them. The objective is to obtain reliable answers to questions which often takes a great deal of persistence and insistence in asking "dumb" questions until the auditor understands how requirements are or are not satisfied.

- *Ability to gain cooperation.* With other auditors, the sponsor, the auditee, and superiors, the world's smallest package is a person wrapped up with himself. Auditors who believe they do not need others are the smallest and most ineffective auditors. Infinite wisdom is not possessed by anyone, and without cooperation with all persons involved in the audit, much will be missed.

- *Ability to reach decisions.* The auditor must be able to separate facts from opinion, compile information and evidence, and compare evidence with standards. This is not instant decision making. The auditor is expected to keep asking questions until an understanding of how requirements are or are not satisfied is reached. Sometimes only persistent *dumb* questions will bring out the truth.

If two people always agree one is not necessary

- *Ability to administer.* Although this is primarily the responsibility of the lead auditor, all auditors should have some administrative ability. Audit results must be documented and retrievable for follow-up, reaudits, and analysis. Reports are usually required of not only individual audit results but an administrative overview of the effectiveness versus the cost is often required.

- *Ability to work independently, systematically, and energetically.* This complements the need for cooperation just mentioned; it is the absence of over-dependency. More often than not, the auditor works alone; even with an audit team, team members each go their separate ways and evaluate different aspects of an operation. A person needs to be task-oriented.

- *Ability to acquire and use special knowledge and skills.* You want to employ knowledgeable individuals in the auditing department, but no matter how experienced and knowledgeable one is the auditing will give them an opportunity to learn a great deal more. An auditor should be an individual with a thirst for knowledge.

- *Ability to adapt to changing work assignments and conditions.* An auditor may be required to work in a foundry, a heavy manufacturing shop, a gage manufacturing shop, an electronic assembly shop, an office environment, etc. In all cases the individual needs to be comfortable and have the ability to put those working in the area at ease.

- *Good outward appearance.* The first impression is a lasting impression, therefore appearance is important — not only to give a good first impression, but to be respectful of the department or company being audited.

- *Intelligent, alert, comprehending, and reasoning.* Some of these characteristics have been mentioned. The operative word is *alert.* If one is not alert all the other characteristics lose their value. To be alert you need to be rested, sober, and genuinely interested in the audit.

- *Emotionally stable, calm.* The auditor is asked to make a judgment concerning the department's or company's quality system. Emotions cannot have an influence in this judgment otherwise irreparable harm could result from actions taken on an unreliable report.

- *Good character.* The auditor is representing a company and is often one of the few contacts the auditee has with that company. The impression made has some affect on the auditee's opinion of the entire company and future dealings with it. Similarly, the company is relying on an individual to be honest, reliable, constructive, helpful, and diplomatic both internally and with the supplier. Auditors with good character are especially important in the defense industry: just one scandal could ruin a company.

- *Good attitude, careful, curious, and open-minded.* There are many ways to accomplish a task and an auditor must be flexible enough to accept different methods and intelligent enough to know if it is acceptable. An auditor must not suffer from the "not-invented-here" or "that's the way we always do it" syndrome. Ideally, an auditor should start with an open mind and positive attitude. Ask an auditor to read the following:

OPPORTUNITY ISNOWHERE

The proper mind-set should read "opportunity is now here"; others may read "opportunity is nowhere."

- *Punctual.* Punctual auditors give the impression that they mean business. A chronic late arriver irritates everyone and wastes valuable time.[7]

- *Good eyesight.* This is commonly required in manufacturing industry and inspection, particularly today, when micro-electronic engineering may involve evaluation of a product that is assembled underneath a microscope. Good eyesight may include not only visual acuity but also color vision; i.e., the ability to distinguish between different colors of electrical wiring may be necessary.[7]

On reviewing these personal traits you quickly perceive that not many people, if anyone, possesses them all. The value of an audit team becomes more apparent now. Each member of the team should complement the other, not necessarily duplicate. The minimum traits that each auditor must possess are good character, emotional stability, and the ability to listen.

You have two ears and one mouth for a reason —
you should listen twice as much as you talk

8.3 Education

Having a formal education does not always mean that a person is adequately educated, nor can you assume that without a formal education a person is not adequately educated. Guidelines, however, must be set. These guidelines may eliminate some people, but in the long run they will pay off by not investing the company's time in several people to gain just one competent auditor.

As a minimum, all auditors should have a high school education. This should assure that a person was given instruction in the *Three Rs:* reading, writing, and arithmetic. Unfortunately, a high school diploma doesn't assure that a person possesses an adequate command of the *Three Rs.*

The best way to determine whether persons possess the basic competency skills is to have them accompany an auditor on an evaluation of an internal process. This evaluation, led by a senior auditor, should be limited in scope so that it only takes a few hours. During the audit the person should be asked to analyze a procedure, make a calculation, and write a summary of evaluation analysis. If you cannot afford this intense an interview you should ask a sufficient amount of questions to determine a person's command of the *Three Rs*.

You might think that this approach is too simplistic and even degrading, but the fact is that industry today has many engineers who obtained their position through seniority. As a consequence, some people holding the title of *engineer* are not able to write a grammatically correct sentence, make simple mathematical calculations, or analyze data.

8.4 Experience

The most desirable situation is to have well-seasoned quality assurance auditors that have worked within organizations with similar products or services. This often is not the case and, therefore, you need to look toward other areas of experiences that are supportive of the auditor's responsibility. Quality, technical, and supervisory experience are key areas to look for. These qualities coupled with the experience of being the auditee will not only technically, but emotionally, complement one's preparation for learning the audit profession.

Until you have walked in their moccasins you will not get willing attention

Quality

The word *auditor* connotes inspection or evaluation. People who have worked in quality assurance planning or engineering are usually familiar with most of the documents identified in Chapter 1. They have probably been challenged on the meaning or interpretation of several areas of these specifications and have had to do much research and soul searching. These experiences not only increase one's knowledge but also one's self-confidence.

To just read the quality documents without the responsibility of implementing, defending, or explaining them is not enough. It takes more than reading to make a person competent; ask any surgeon.

Technical

Auditors are often "led down a primrose path" if they are not knowledgeable of the technical aspects germane to the area being audited. This is not to be interpreted as requiring a technical degree for all auditors: a person with a degree in turbine engine design may not be capable of evaluating a machine shop, whereas a master machinist with a high school education probably would be. Auditors may find it necessary to request the assistance of a technical expert at times, but generally speaking the auditor should have practical experience in the area being evaluated.

Supervision

People who have been in supervision have had some on-the-job training in dealing with a variety of personalities. They understand the benefits of teamwork and the damage caused by an adversarial relationship. They also appreciate the importance of reliable and verifiable information.

8.5   Education/Experience Guideline

You should not treat a person's education and experience background separately. Additionally, auditing experience at another company may not meet your requirements. Table 8.1 shows a matrix that has been used successfully by some companies.

Education	H.S./GED	H.S. + 2 yrs/CQE	H.S. + 4 yrs
QA Experience No. of Audits*	3 years 2	2 years 2	6 months 2

*These audits are used to evaluate the auditor. The first audit should have the candidate observe a qualified auditor. The second audit should be performed by the candidate with an experienced auditor observing. Based on these audits a determination can usually be made: fully qualified, needs more training, not qualified. This applies to all auditors new to an organization, regardless of previous auditing experience.

Table 8.1   Minimum Auditor Education and Experience

For experienced auditors this is an important time of orienting them to an organization's operation. For a person being considered for promotion into an auditing department this should be explained during the interview as a prerequisite for obtaining the position.

8.6 Putting It All Together

It is highly unlikely, if not impossible, to find all of these qualifications (independence, personal traits, education, and experience) in a single person. An audit team, on the other hand, may possess the essential ones for a particular audit.

Be careful when you meet people that know a lot and think they know it all, because they actually know little.

The smallest package in the world is a person wrapped up in himself

CHAPTER 9

PRODUCT AUDITS

Product auditing is an analysis of elements of a process and appraisal of completeness, correctness of conditions, and probable effectiveness.[1] Its purpose is one of assurance rather than control and its perspective is from the user's viewpoint.

The user in this case is not the DCAS, Air Force, Army, or Navy Resident Quality Assurance Representative, but rather the command that actually uses and/or maintains the product in performance of their assignment/responsibility.

9.1 Audit Scope

When evaluating fitness of items for use we normally think of final inspection of the interface characteristics followed by a functional test that closely replicates the item's use. This is normally performed on any system the government purchases and is only part of a product audit.

The scope of the audit is related directly to identifying standards against which the product's fitness will be measured. The item being audited has been fully accepted and, therefore, not every characteristic needs to be audited; your internal quality program is designed to verify the product's acceptance. The product audit plan should encompass performance, interface, maintenance, and interchangeability characteristics: areas that the user is most interested in.

If a diode is the end product, the plan will be simple; however an airplane will require a complex audit plan. A product audit of a large complex system such as an airplane cannot be solely conducted just before shipment. There are many components and subassemblies that become effectively encapsulated in the end item, and therefore need to be audited as if they were end items in and of themselves.

9.2 Audit Frequency

To establish a frequency you must consider the production volume, criticality, complexity, and maturity of the item.

- *Production volume.* Daily audits will be scheduled for high production products with short production cycles, in recognition of rapid quality changes that can take place under high-volume conditions. Weekly audits will be scheduled for products of medium production volume. Monthly audits may be required for long production cycle products. Airplanes and ships, on the other hand, may be done only once as subsystems and/or during the government's Physical Configuration Audit (PCA).

- *Criticality.* During the design phase, engineering may identify a limited number of characteristics that must be audited on each item as a safety/reliability measure.

- *Complexity.* As products become increasingly complex, the product auditing is more often conducted at several stages: after final inspection, after packaging, upon receipt, and/or during actual performance by the government. The frequency is varied based on the stage. The bulk of the characteristics may be evaluated at the most economical stage.

  - Inspected — most economical, but does not reflect effect of packing, shipping, storage, or usage.

  - Packaged — requires unpacking and repacking, but evaluates effect of original packing.

  - Received — difficult to administer, but evaluates the effects of storage, shipment, and unpacking.

  - Performance — the ideal, but also the most difficult to administer.

  - Maturity — new products may receive twice the frequency of audits as mature ones that have been through the initial production fine-tuning.

In the final analysis, your quality history on a product (history of similar products for new programs) will dictate the necessary audit frequency.

9.3 Quality Rating

Again, the product audit has the viewpoint of the user, and therefore the product audit quality rating must also reflect the interest of the user. Several rating methods are used today. One of the methods that is particularly effective and gains the attention of management weights each characteristic with a *seriousness classification.*

The scope of the audit (9.1) also identifies the standards or characteristics that will be evaluated. Using the *seriousness classification* method, each characteristic is coded to the classifications that follow. The sum of the actual demerits divided by the possible demerits will yield a percent defective for each classification. Any defects in classifications A, B, or C obviously are unacceptable and must receive prompt attention.

Seriousness Classification (6)

Class A — Critical (Demerit Value, see NOTE)

a. Will surely cause a failure that could injure personnel or jeopardize a vital agency mission.

NOTE: Critical items must be investigated immediately and the necessary agencies notified as soon as the deficiency has been confirmed. One critical characteristic found discrepant will cause a product audit to fail.

Class B — Very Serious (Demerit Value, 100)

a. Will surely cause an operating failure of the unit in service which cannot be readily corrected in the field, e.g., open relay winding; or

b. Will surely cause intermittent operating trouble, difficult to locate in the field, e.g., loose connection; or

c. Will render unit totally unfit for service, e.g., control dial does not return to normal after operation.

Class C — Serious (Demerit Value, 50)

a. Will probably cause an operating failure of the unit in service which cannot be readily corrected in the field, e.g., protective finish missing from coaxial plug; or

b. Will surely cause an operating failure of the unit in service which can be readily corrected in the field, e.g., relay contact does not make; or

c. Will surely cause trouble of a nature less serious than an operating failure, such as substandard performance, e.g., protrector block does not operate at specified voltage; or

d. Will surely involve increased maintenance or decreased life, e.g., single contact disk missing; or

e. Will cause a major increase in installation effort by the user, e.g., mounting holes in wrong location; or

f. Defects in appearance or finish that are extreme in intensity, e.g., damaged to a degree that requires complete refinishing.

Class D — Moderately Serious (Demerit Value, 10)

a. May possibly cause an operating failure of the unit in service, e.g., contact less than minimum; or

b. Likely to cause trouble of a nature less serious than an operating failure, such as substandard performance, e.g., signal does not operate within specified limits; or

c. Likely to involve increased maintenance or decreased life, e.g., dirty contact; or

d. Will cause a minor increase in installation effort by the user, e.g., mounting bracket distorted; or

e. Major defects in appearance, finish, or workmanship, e.g., finish conspicuously scratched, identification omitted, or illegible.

Class E — Not Serious (Demerit Value, 1)

a. Will not affect operation, maintenance, or life of the unit in service (including minor deviations from engineering requirements), e.g., sleeving too short; or

b. Minor defects of appearance, finish, or workmanship, e.g., slightly scratched finish.

This rating system (Table 9.1) may be used to identify serious deficiencies that need immediate attention, evaluate quality trends of the product, and compare a specific product to the history of a similar family of products. To compare one product to another the characteristics and the demerit scheme must be similar.

Defect Class	Potential Demerits	Actual Demerits	Percentage Acceptable
A	0 x fail	n/a	n/a
B	4 x 100 = 400	0 x 100 = 0	100
C	3 x 50 = 150	1 x 50 = 50	67
D	14 x 10 = 140	4 x 10 = 40	71
E	20 x 1 = 20	3 x 1 = 3	85

Table 9.1 Rating System

CHAPTER 10

SPECIAL PROCESS AUDITS

The definition of *special process* is often misunderstood. For the purpose of a common understanding the following definition will be assumed.

> A particular method of doing something, generally involving a number of steps or operations in a specific sequence which affects an article in a manner which cannot economically be determined, verified, or controlled adequately by inspection upon completion.[8]

Some examples of special processes include:

- Anodic, cathodic, and immersion coatings
- Spray application of surface coatings
- Thermal treatment of materials
- Joining of materials
- Potting, encapsulation, and impregnation
- Mixing and blending of compounds
- Nondestructive testing
- Environmental testing
- Chemical cleaning

10.1 Similarity to Quality System Audits

As with a quality system audit you need to:

- Know the specification.
- Know the reason for and objective of the audit.
- Establish an audit policy and protocol.
- Develop a checklist.
- Issue a report.

- Obtain corrective action.

- Perform follow-up audits.

- Establish qualification criteria for the auditor.

Most of these steps are so similar to the quality system audit that they need not be expanded. The audit policy and protocol, checklist, and auditor qualification, however, are sufficiently different that further discussion is necessary.

10.2   Audit Policy and Protocol

Special processes encompass operations that are completed in seconds to those that take days. They also vary from high volume to one or two items. To establish a detailed policy and protocol that would be applicable to all special processes would be impossible.

The policy and protocol need to be consistent in establishing what to audit and how to determine the frequency of the audit. They must be definitive enough to consistently follow, while being flexible enough to apply to all special processes.

An interdepartmental committee should guide the preparation of this policy and protocol to avoid the risk of the system reflecting one department's outlook.[9] The first item this committee should address is an agreed upon listing of what comprises a special process. A listing should be generated specifically identifying each process that is applicable to a company or its suppliers.

Once you have defined special processes you need to review each standard to find out any special audit frequency that may be required. Chapter 4 provides general guidelines when a particular frequency is not required.

10.3   Worksheets

Many large companies have translated industry and military specifications into a company-controlled document. Although this may clarify some points it also adds to confusion, especially with the supplier that performs basically the same process for several customers, all of which have their own peculiar vintage of a common specification.

If at all possible you should comply with the actual industry/military specification with any special requirements identified on the drawing or a supplementary document. You will then be able to develop your worksheet to the same specification that the government has imposed.

This will yield several tangible and intangible benefits:

1. You will be better able to prove your contractual compliance which, in most cases, calls out the industry or military specification.

2. Your government representative will be better able to understand your program. You will not have to spend the time explaining how your specification *meets or exceeds* the contractual obligation.

3. Your supplier will not have one more specification to interpret. The industry or military specification is, in the end, what they must comply with and what they are most familiar with. Something is always lost in the interpretation.

A single worksheet for each specification will allow you to use it internally as well as at your suppliers. Using these common criteria you will be able to trend internal and external compliance and take the necessary action to maintain a highly reliable special process activity.

When conducting a special process audit it is not always limited to conducting an on-site audit. Often a sample of the process needs to be sent to a laboratory to conduct a physical and chemical analysis. This needs to be considered when developing the worksheet so that the auditor knows to obtain a sample before leaving the area.

10.4 Auditor Qualifications

Ideally a person conducting a special process audit should be fully qualified as an auditor in accordance with the guidelines given in Chapter 8 and have a college degree in the scientific aspects of the process. Realistically, few companies can afford this luxury; they may have a few technical experts in the audit department, but these experts cannot possibly have sufficient knowledge to audit every process.

When you are requested to audit a process for which the audit department does not have sufficient expertise you need to look elsewhere. Your first choice should be to solicit help from your engineering department. A fully qualified auditor assisted by a technically competent engineer make a credible team. An auditor who participates in several of these team audits will eventually become fully competent to perform the audit singularly.

This approach has two benefits in addition to the value of the audit.

1. The engineers who assist the auditors obtain on-the-floor experience that will enable them to contribute more profitably and, in some cases, realistically to the processing operation.

2. The auditor's personal knowledge is expanded, which will strengthen the capability, efficiency, and credibility of the audit department.

CHAPTER 11

CORPORATE AUDITS

Corporate audits connote that an executive body has elected to evaluate the effectiveness of the companies for which they are responsible. Although the guidelines are similar to an internal or supplier audit, the benefits are greatly expanded.

11.1 Single Audits

The corporation I work for has many divisions scattered throughout the United States and several foreign countries. There are many interdivisional orders which necessitate including sister divisions on each approved supplier list.

This in turn requires approvals/audits of sister divisions multiplied and duplicated many times over. The corporate audit is a unified approach that allows each division within the corporation to use this single audit as its approval.

Many hundreds and even thousands of dollars are saved in the avoidance of duplicate audits that require planning, travel expenses, living expenses, and time. Additionally, using the single audit avoids multiple disruptions to the division being audited.

The audit policy, procedure, criteria, and schedule need to be mutually agreed upon by all divisions to assure that they meet their internal and contractual requirements.

11.2 Information Exchange

The lead auditor needs to be fully qualified in accordance with the guidelines given in Chapter 8; however, the remaining members of the audit team may be functional engineers/managers.

Audit team selection is geared to having functional specialists/managers with similar responsibilities evaluate the equivalent of their department. Not only will they have the special knowledge, they also are able to establish contacts for future help or assistance.

I have conducted several of these audits and, without exception, the auditors and auditees have taken with them new approaches to either adopt or avoid and new professional contacts with whom to share ideas. This benefit alone has helped keep our corporation current with industry practices and, in several cases, allowed us to tread new ground.

11.3 Needs and Strengths

The corporation leading all the audits allows for a better understanding of each of the division's strengths and weaknesses. In a sense the corporation is a service organization, and without this objective insight the corporation's support cannot be as effective. The executives often see their organization through rose-colored glasses because they are shielded from the everyday operation. The corporate audit, however, does not have this buffer from the everyday, real world, operation.

The audit results objectively indicate where a division's strengths and weaknesses are. With this knowledge the corporation is able to exchange information and, in some cases, resources from division to division in an effort to bring the total corporation up to a higher standard.

The corporation is also able to identify the most talented expert in a variety of disciplines. A list can then be generated to identify these experts. When information or consultation is needed these recognized experts are readily accessible. It is safer and more cost effective to consult with a known expert than to take your chances with an unknown consultant.

11.4 Overview of Corporate Audits

Purpose of a Corporate Audit

- To support and help a division — not to criticize a division.
- To open lines of communication within the corporation.
- To verify compliance to the corporate and divisional quality assurance policies.

Organization

- Corporate quality assurance leads audit.
- Functional specialists/managers from other divisions of the corporation act as auditors to:
  - Open lines of communication.

– Share technology.
– Explain how successful improvements have been made at other divisions.
– Have an objective understanding of the division being audited.

Basis for a Corporate Audit Form

- Corporate quality assurance policy.

- Common industry standards.

- Processes that are common among divisions.

- Input from corporate quality council (quality managers of each division).

- Divisional quality assurance manual.

- Previous audit results/corrective action.

The Audit

- Review divisional procedures.

- In-brief.

- Perform audit.

- Daily out-brief with team leader and division quality assurance management.

- Exit interview.

- Document results and suggestions in a letter to corporate and divisional quality assurance.

- Division management has:
  – Objective audit of their compliance to quality assurance requirements — without threat.
  – Suggestions for improvements that have been successful at other divisions.
  – New contacts for future quality assurance support.

- Divisions participating on the audit team have:
  – Insight into different approaches to adopt, modify, or avoid.
  – New contacts for future quality assurance support.

- Corporate management has:
  – Insight as to how they can best support each division.
  – Knowledge of where technical strengths are so that they may be used/ transferred to other divisions.
  – Assurance that the divisions are harmoniously maintaining a dynamic quality assurance organization.
  – Data base from which other corporate quality assurance activities are germinated to better unify, educate, or support the divisions.

11.5   Unified Supplier Base Audits

Corporations with multiple divisions often use many of the same suppliers. Many suppliers are from the immediate community, but defense programs require highly sophisticated and/or specialized suppliers that are not always located next door.

There is a major corporation that has analyzed the suppliers used by six divisions located in Canada and four states. They found that 38% of their suppliers were used by more than one division. They also discovered that occasionally auditors from different divisions and states were staying at the same hotel at the same time and auditing suppliers that were just blocks apart. Both auditors had to travel more then 2,000 miles, rent a car, check into a hotel, etc. This was determined to be not cost effective.

The corporation executives decided to unify their approach to supplier control. In order to do this they recognized that they had to develop a common audit policy and procedure (Chapter 4), audit worksheet (Chapter 5), audit reporting (Chapter 6), corrective action and follow-up requirement (Chapter 7), and auditor qualification (Chapter 8) in order to reliably use each others audit results. This would eliminate the need for each division to separately audit a supplier that was common to them. Additionally, by identifying geographic concentrations of suppliers they would be able to place field representatives in strategic areas to save travel time and money.

CHAPTER 12

ACCEPT/REJECT CRITERIA FOR A QUALITY SYSTEM AUDIT PROGRAM

MIL-Q-9858A, paragraph 6.2, states: "Criteria for approval and rejection shall be provided for all inspection of product and monitoring of methods, equipment, and personnel. Means for identifying approved and rejected product shall be provided."

We also should establish accept/reject criteria for our quality system audit programs. There are too many "programs" being introduced that are not evaluated as to their effectiveness, and we can little afford any more.

12.1   The Evaluator(s)

Many people would say that the audit department, being closest to the actual audits, should establish the accept/reject criteria for the audit program. That's like saying the acceptance or rejection of a new product should be determined by the inventor or the manufacturer.

My son-in-law and I are constantly coming up with new inventions that we believe would make us a fortune. Our wives are perceptive enough to know that our "great ideas" would drive us to the poor farm. Many people have thought of "great inventions," invested a lot of money to produce and market them, and subsequently lost their shirts. The consumer or customer did not share their opinion. The same is true of an audit program: In the final analysis, the customer is the one to decide if the program is worthwhile.

12.1.1   The Customer

The customer is not only the government: it is the next person/department to use the service or audit results. The list of customers includes, but is not limited to:

- Contracts

- Material control
- Purchasing
- Manufacturing
- Engineering
- Quality
- Government

All of these departments rely on the quality program to identify weaknesses before they get out of hand. The most immediate customers of an audit program are the departments within a company, not the agency that purchases your product or service.

12.2   Accept/Reject Criteria

The audit department may be able to evaluate its program in some areas, but the chief definer of what is acceptable and what is not is the customer.

The audit department is a support function or a service organization that needs to be sensitive to business needs. To understand if the audit program is acceptable or not you need to ask your customers if their needs are being met.

If the audit department is surprised by any of the following items then the audit program is not effective:

- Applicable contract requirements are not consistently and effectively being communicated to all departments.
- Engineering drawings and specifications are not being maintained current in the planning department or on the manufacturing floor.
- Purchasing is not keeping the supplier base informed of changes that need to be incorporated on current purchase orders.
- Sampling inspection plans are not being followed consistently.
- The certification records for magnetic particle inspectors do not contain the required information.

This is just a sample of the many areas that an effective audit program should bring to the attention of its "customers" so that they are able to make the necessary corrections. If they are surprised then the program is not effective.

The customer — the evaluator of an acceptable quality systems audit

I HEAR (about audits) I FORGET

I SEE (read this book) I REMEMBER

I DO (perform an audit) I UNDERSTAND

CHAPTER 13

TERMS AND DEFINITIONS

The following is a compilation of terms and definitions that may be used within the auditing discipline. There are numerous standards published today from which I selected the most appropriate definition.

Approval

An act of endorsing or adding positive authorization, or both.[10]

Appraisal

An estimate or determination of the significance, importance, or value of something.

Assessment

See **Appraisal.**

Audit

A planned, independent, and documented assessment to determine whether agreed upon requirements are being met.

Audit Finding

The determination and recording of adequacy or inadequacy of conformance of a product, process, or procedure characteristic to the specified standard.[6]

Audit Program

The organizational structure, commitment, and documented methods used to plan and perform audits.

Audit Standard

The official description of essential characteristics of audits which reflects on that and similar products. The plan should identify not only the action to be taken but who is responsible for that action and by what date the preventive action will be in effect.

Audit Team

The group of individuals conducting an audit under the direction of a team leader.[6]

Auditee

The organization to be audited.[11]

Auditing Organization

The unit or function that carries out audits through its employees. This organization may be a department of the auditee, client, or an independent third party.[11]

Auditor

The individual who carries out the audit. The auditor is appointed by the auditing organization selected by the client or by the client directly.[11]

Capability

Ability of an entity to perform designated activities.[2]

Certificate of Conformance

A document signed by an authorized party affirming that a product or service has met the requirements of the relevant specifications, contract, or regulation.[1]

Certificate of Compliance

A document signed by an authorized party affirming that the supplier of a product or service has met the requirements of the relevant specifications, contract, or regulation.[1]

Certification

The procedure and action by a duly authorized body of determining, verifying, and attesting in writing to the qualifications of personnel, processes, procedures, or items in accordance with applicable requirements.[1]

Certified Test Report

A written and signed document, approved by a qualified party, that contains sufficient data and information to verify the actual properties of items and the actual results of all required tests.[10]

Characteristic

Any distinct property or attribute of an item, process, or service that can be described and measured to determine conformance or nonconformance to specified requirements.[6]

Checks

The tests, measurements, verifications, or controls placed on an activity by means of investigations, comparisons, or examinations, to determine satisfactory condition, accuracy, safety, or performance.[10]

Cleanness

A state of being clean in accordance with predetermined standards, and usually implies freedom from dirt, scale, heavy rust, oil, or other contaminating impurities.[10]

Client

The person or organization requesting the audit. Depending upon the circumstances, the client may be the auditing organization, the auditees, or a third party.[11]

Compliance

An affirmative indication or judgment that the supplier of a product or service has met the requirements of the relevant specifications, contract, or regulation; also the state of meeting the requirements.[1]

Confirmation

That representations (verbal or written) are in accordance with data or findings obtained from different sources.[6]

Conformance

An affirmative indication or judgment that a product or service has met the requirements of the relevant specifications, contract, or regulation; also the state of meeting the requirements.[1]

Conformity

The fulfilling by an item or service of specification requirements.[1]

Contractor

Any organization under contract for furnishing items or services. It includes the terms vendor, supplier, subcontractor, fabricator, and subtler levels of these where appropriate.[10]

Corrective Action

Action taken to eliminate the causes of an existing undesirable deviation or nonconformity to prevent recurrence.[2]

Defect

The nonfulfillment of intended usage requirements.[2]

Defective Material

A material or component which has one or more characteristics that do not comply with specified requirements.[10]

Design Review

A formal, documented, comprehensive, and systematic examination of a design to evaluate the design requirements and the capability of the design to meet these requirements and to identify problems and propose solutions.[12]

Deviation

A nonconformance or departure of a characteristic from specified requirements.[10]

Deviation Permit

Written authorization, before production or provision of a service, to depart from specified requirements for a specified quantity or for a specified time.[1]

Documented

Any written or pictorial information describing, defining, specifying, reporting, or certifying activities, requirements, procedures, or results.[10]

Evaluation

A systematic examination of the capability of an organization, or part thereof, to meet given requirements.[2]

Examination

An element of inspection consisting of investigation of materials, components, supplies, or services to determine conformance to those specified requirements which can be determined by such investigation. Examination is usually nondestructive and includes simple physical manipulation, gaging, and measurement.[10]

Finding

A conclusion of importance based upon observation(s).

Follow-Up Audit

An audit whose purpose and scope are limited to verifying that corrective action has been accomplished as scheduled, and determining that the action was effective in preventing recurrence.

Grade

An indicator of category or rank related to features or characteristics that cover different sets of needs for products or services intended for the same functional use.[2]

Guidelines

Particular provisions which are considered good practice but which are not mandatory in programs intended to comply with this standard. The term *should* denotes a guideline; the term *shall* denotes a mandatory requirement.[10]

Handling

An act of physically moving items by hand or mechanical means, but not including transport modes.[10]

Improvement

The attainment of a new level of performance that is superior to any previous level.[10]

Independence

Freedom from bias and external influences (e.g., an auditor would not be considered independent if he performed an audit of a former employer for a third party).[6]

Inspection

Activities, such as measuring, examining, testing, or gaging one or more characteristics of a product or service, and comparing these with specified requirements to determine conformity.[1, 2]

Inspector (Owner's or Installer's)

A qualified inspector employed by the owner or installer whose duties include the verification of quality related activities or installations, or both.[10]

Inspector (State or Code)

A qualified inspector employed by a legally constituted agency of a municipality or state of the United States, or Canadian Province, or regularly employed by an authorized inspection agency and having authorized jurisdiction at the site of manufacture or installation.[10]

Item

Any level of unit assembly, including structure, system, subsystem, subassembly, component, part, or material.[10]

Lead Auditor

An auditor who supervises other auditors during an audit as a team leader.[11]

Material

A substance or combination of substances forming components, parts, pieces, and equipment items (including machinery, castings, liquids, formed steel shapes, aggregates, cement, etc.)[10]

Monitoring

See **Quality Surveillance.**

Nonconformance

The nonfulfillment of specified requirements. This covers the departure or absence of one or more quality characteristic or quality system element from specified requirements.[2, 12]

Objective Evidence

Qualitative or quantitative information, records, or statements of fact pertaining to the quality of an item or service or to the existence and implementation of a quality system element, which is based on observation, measurement, or test and which can be verified.[12]

Observation

A conclusion and statement of fact made during an audit and substantiated by objective evidence.[11]

Part

An item which has work performed on it and which is attached to and becomes part of a component before completion of a component.[10]

Physical Audit Activities

Those activities by which information is obtained to verify the auditee's compliance with the applicable standards (e.g., interviews, observations, examination of evidence, measurements, and tests).[6]

Planning (for Quality)

Establishing and developing the objectives and requirements for quality of an entity and the managerial and operational procedures for attainment.[2]

Procedure

A document that specifies the way to perform an activity.[2]

Process Quality Audit

An analysis of elements of a process and appraisal of completeness, correctness of conditions, and probable effectiveness.[1]

Procedure

A document that specifies or describes how an activity is to be performed. It may include methods to be employed, equipment or materials to be used, and sequence of operations.[10]

Procurement Documents

Contractually binding documents that identify and define the requirements that items or services must meet in order to be considered acceptable by the purchaser.[10]

Product Liability (Service Liability)

A generic term used to describe the onus on a producer or others to make restitution for loss related to personal injury, property damage, or other harm caused by a product or service.[2]

Product Quality Audit

A quantitative assessment of conformance to required product characteristics.[1, 6]

Purchaser

The organization or organizations responsible for issuance of administration of contract, subcontract, or purchase order.[10]

Qualification

The status given to an entity when the fulfillment of specified requirements has been demonstrated or the process of obtaining that status.[2]

Quality

The totality of features and characteristics of a product or service that bear on its ability to satisfy stated or implied needs now and in the future.[1]

Quality Assurance

All those planned or systematic actions necessary to provide adequate confidence that a product or service will satisfy given requirements for quality.[13]

Quality Audit

A systematic and independent examination to determine whether quality activities and related results comply with planned arrangements and whether these arrangements are implemented effectively and are suitable to achieve objectives.[1, 2, 12]

Quality Control

The operational techniques and the activities used to fulfill requirements of quality.[1, 2, 13]

Quality Engineering

That branch of engineering which deals with the principles and practice of product and service quality assurance and control.[1]

Quality Look/Quality Spiral

Conceptual model of interacting activities that influence the quality of a product or service in the various stages ranging from the identification of needs to the assessment of whether these needs have been satisfied.[1]

Quality Loop (Quality Spiral)

Conceptual model of interacting activities that influence the quality of a product or service in the various stages ranging from the identification of needs to the assessment of whether these needs have been satisfied.[2]

Quality Management

That aspect of the overall management function that determines and implements the quality policy.[1, 2]

Quality Manual

A document setting out the quality policies, systems, and practices of an organization.[2]

Quality Measure

A quantitative measure of the features and characteristics of a product or service.[1]

Quality Plan

A document setting out the specific quality practices, resources, and activities relevant to a particular product, process, service, contract, or project.[1, 2]

Quality Policy

The overall quality intentions and direction of an organization as regards quality as formally expressed by top management.[1, 2]

Quality Program

The documented plans, organizational structure, and activities that are implemented to control the conformance of a product or service to specified requirements and to provide evidence of such conformance.[6]

Quality Program Audit

The documented activity performed to verify, by examination and evaluation of objective evidence, that applicable elements of the quality program have been developed, documented, and implemented effectively, in accordance with specified requirements.[6]

Quality Surveillance

The continuing monitoring and verification of the status of procedures, methods, conditions, products, processes, and services, and analysis of records in relation to stated references to ensure that requirements for quality are being met.[1, 2]

Quality System

The organizational structure, responsibilities, procedures, processes, and resources for implementing quality management.[1, 2, 12, 13]

Quality System Audit (Quality Plan Audit)

A documented activity performed to verify, by examination and evaluation of objective evidence, that applicable elements of a quality system are suitable and have been developed, documented, and implemented effectively in accordance with specified requirements.[1, 11]

Quality System Review

A formal evaluation by management of the status and adequacy of the quality system in relation to quality policy and/or new objectives resulting from changing circumstances.[1, 2]

Quality System Survey

An activity similar to a quality system audit but having a different purpose. A survey takes place with relation to a prospective procurement of a service or an item. Generally, it is conducted before contract award and is used to evaluate the overall capability of a prospective supplier/contractor including the adequacy and implementation of the supplier/contractor quality assurance program.[14]

Relative Quality

Degree of excellence of a product or service.[1]

Reliability

The ability of an item to perform a required function under stated conditions for a stated period of time.[2]

Reliability, Numerical

The probability that an item will perform a required function under stated conditions for a stated period of time.[1]

Reliability Engineering

That engineering function dealing with the principles and practices related to the design, specification, assessment, and achievement of product or system reliability requirements and involving aspects of prediction, evaluation, production, and demonstration.[1]

Repair

The process of restoring a nonconforming characteristic to a condition such that the capability of an item to function reliably and safely is unimpaired, even though that item still may not conform to the original requirement.[10]

Report

Something (document) that gives information for record purposes.[10]

Rework

The process by which a nonconforming item is made to conform to a prior specified requirement by completion, remachining, reassembling, or other corrective means.[10]

Root Cause

A fundamental deficiency which results in a nonconformance and must be corrected to prevent recurrence of the same or similar nonconformance.

Source Surveillance

A review, observation, or inspection for the purpose of verifying that an action has been accomplished as specified at the location of material procurement or manufacture.[10]

Special Process Audit

A particular method of doing something, generally involving a number of steps or operations in a specific sequence which affects an article in a manner which cannot economically be determined, verified, or controlled adequately by inspection upon completion.[8]

Specification

The document that prescribes the requirements with which the product or service has to conform.[1, 2]

Standard

The result of a particular standardization effort approved by a recognized authority.[10]

Statistical Process Control

The application of statistical techniques to the control of processes.[1]

Statistical Quality Control

The application of statistical techniques to the control of quality.[1]

Supplier

Processors or manufacturers of materials and components who sell not to consumers but mainly to other manufacturers for further processing, or customers who embody the supplier's products into a broad service system.[13]

Survey

An examination for some specific purpose; to inspect or consider carefully; to review in detail. (Some authorities use the words *audit* and *survey* interchangeably. The word *audit* implies the existence of some agreed upon criteria against which the plans and execution can be checked. The word *survey* implies the inclusion of matters not covered by agreed upon criteria.)

Testing

A means of determining the capability of an item to meet specified requirements by subjecting the item to a set of physical, chemical, environmental, or operating actions and conditions.[1]

Traceability

The ability to trace the history, application, or location of an item or activity and like items or activities by means of recorded identification.[1, 2]

Use-As-Is

A disposition which may be imposed for a nonconformance when it can be established that the discrepancy will result in no adverse conditions and that the item under consideration will continue to meet all engineering functional requirements including performance, maintainability, fit, and safety.[10]

Verification

The act of reviewing, inspecting, testing, checking, auditing, or otherwise establishing and documenting whether items, processes, services, or documents conform to specified requirements.[1]

Waiver (Concession)

Written authorization to use or release a quantity of material, components, or stores already manufactured but not conforming to the specified requirements.[1, 2]

Working Papers

All documents required by the auditor or audit team to plan and implement the audit. They include audit schedules, auditor assignments, checklists and reporting forms used by auditors, meeting agenda and minutes, etc.[6]

APPENDIX A

MIL-Q-9858A/AQAP-1 AUDIT WORKSHEET

Worksheet Instructions

The following information will clarify the use of this worksheet:

Codes

1 — Identify the written procedure. A procedure is a document that specifies or describes how an activity is to be performed.

2 — Identify the instruction. An instruction is an order or direction outlining a sequence of steps to follow in performing a specific inspection, test, or work function.

3 — Verify documentation. Documentation is anything printed or written that is relied upon to prove or support, e.g., purchase order, forms, flow charts, records, travelers, drawings, memos, etc. A written explanation for a uniform application and preparation of documentation may be required.

4 — A minimum of three observations shall be performed.

Procedure/Document Reviewed

When Codes 1 or 2 are identified the auditee is to record the applicable procedure/instruction in this column. This is to be done before the survey/audit.

Number of Observations Made

When Code 4 is identified the auditor is to record the actual number of observations made in this column.

Compliance — Yes/No

After evaluation of each applicable question the auditor is to identify whether or not the auditee is in compliance.

N/A (✓)

This column is to be checked (✓) if the question is not applicable to the auditee.

Remarks (✓) See Below

This column is to be checked (✓) if further explanation is necessary. The explanation is to be given at the end of the applicable section and identified to the question number.

Specification/Worksheet Cross-Reference

MIL-Q-9858A		AQAP-1	
3.1	A2, A3, L9	201	A10, B14, E1, E2, E9, G1, H5, H6
3.2	A5–A8	202	A1, A2
3.3	A1, A9, A10, E1, E2, H1, I1	203	A3, A5, B15
3.4	A11–A15, E15, F6, G4, H12, I7, J6	204	A6–A8
		205a	A1
3.5	L1–L9	205b	A9
3.6	M1–M4, M6	205c	H1
4.1	B2–B7, B9–B16	205d	A11, A13, A14, A16, F6, H12, I7, J6
4.2	-C-		
4.3	-C-	206	A12, L1–L6, L9
4.4	A20	207	B1–B5, B9, B10, B13, I5
4.5	C35	208a	B6, B7, B11, B12, E16
5.1	E3–E15, F6	208b	B1, B8
5.2	E16, E17	209	-C-
6.1	F1–F5, G5	210a	E4, E6–E9, E17, F6
6.2	A10, H2–H10, H12	210b	E10, E11, E16
6.3	I2–I7	210c	E3, E5, E12–E15, F1–F5, I1
6.4	G1–G4, H13, J1–J4, J6	210d	E6, H12, I7, J6
6.5	K1–K3, M5	211	H1
6.6	D1, D2	211a	H1, H3–H5, H7–H10
6.7	A17, A19, F7, G6, H11, J5	211b	H1, H2
7.1	E18, E19	212	F3
7.2.1	N1	213a	H1
7.2.2	N2–N4	213b	I1–I3
7.2.3	N5–N8	213c	I5
		214	D1–D3
		215	K1–K3
		216	A17, A18, F6, F7, G6, H11, H12, I7, J5, J6
		217	A17, G5, J3
		217a	H1, H13, J6
		217b	G1–G3, G6
		217c	J1, J2, J6
		218	A20

Specification/Worksheet Cross-Reference

Survey Category "C"
(Calibration)

MIL-STD-45662A	AQAP-6
5.1 C2, C3	201 C1, C34
5.2 C1, C4, C25	202 C36
5.3 C5, C7, C8	203 C4, C35
5.4 C9–C11	204 C14
5.5 C12, C13, C15, C16	205 C1–C3, C12, C15, C16
5.6.1 C18, C19	206 C13, C27, C28, C30, C31
5.6.2 C20, C21	207 C29
5.7.1 C22–C24	208 C17
5.7.2 C22	209 C9–C11, C18, C26
5.8 C26–C28	210 C19, C20
5.9 C29–C31	211 C24, C25
5.10 C24	212 C32, C33
5.11 C32	213 C22, C23
	214 C14, C15
	215 C5–C8
	216 A20
	217 A20

A. QUALITY PROGRAM	CODE	PROCEDURE/ DOCUMENT REVIEWED	NO. OF OBS MADE	COMPLIANCE		N/A (√)	REMARKS (√) SEE BELOW
				YES	NO		
1. Is management for quality clearly prescribed in written procedures/instruction/policies? (3.3/202, 205a)	1						
2. Do personnel performing quality functions have: (3.1/202)							
a. Sufficient, well defined responsibility?	1						
b. Authority and the organizational freedom to identify and evaluate quality problems?	1						
c. Authority to initiate, recommend, or provide solutions for quality problems?	1						
3. Does Management regularly review the status and adequacy of the quality program? (3.1/203)	1						
4. Is a single standard quality program maintained?	4						
5. Is the contract reviewed to identify and make timely provision for the following special items to assure product quality: (3.2/203)	1						
a. Controls?	1						
b. Processes?	1						
c. Test Equipment?	1						
d. Fixtures?	1						
e. Tooling?	1						
f. Skills?	1						
6. Does the initial quality planning recognize the need and provide for research, when necessary, to update: (3.2/204)	1						
a. Inspection and testing techniques?							
b. Instrumentation?	1						
c. The correlation of inspection and test results?	1						
7. Does the planning provide for review and action to assure compatibility of manufacturing, inspection, testing, and documentation? (3.2/204)	1						
8. Is initial quality planning performed during the earliest practical phase of contract performance? (3.2/204)	2						
9. Is the preparation and maintenance of work instructions monitored? (3.3/205b)	1						

A. QUALITY PROGRAM (cont.)	CODE	PROCEDURE/ DOCUMENT REVIEWED	NO. OF OBS MADE	COMPLIANCE		N/A (/)	REMARKS (/) SEE BELOW
				YES	NO		
10. Is compliance with work instructions monitored? (3.3, 6.2/201)	1						
11. Do records maintained for monitoring work performance indicate the acceptability of the work? (3.4/205d)	3						
12. Do records for monitoring work performance indicate action taken in connection with deficiencies? (3.4/206)	3						
13. Does the quality program provide for the analysis and use of records as a basis for management action (3.4/205d)	2						
14. Are records or data of all essential activities of the quality program maintained? (3.4/205d)	3						
15. Are copies of the contractor's records furnished to the government representative on request? (3.4)	4						
16. Are all contractor quality records available for review by the government representative? (/205d)	4						
17. Is a positive system maintained for identifying the inspection status of supplies? (6.7/216, 217)	2						
18. Are inspection stamps, if used, traceable to the individual responsible for its use? (/216)	3						
19. If inspection stamps are used, is the design distinctly different from government identification? (6.7/)	4						
20. Is assistance in the form of access, documentation, and equipment made available to the government when necessary for verifying contract compliance? (4.4/218)	4						

B. DRAWINGS, DOCUMENTATION, AND CHANGES	CODE	PROCEDURE/ DOCUMENT REVIEWED	NO. OF OBS MADE	COMPLIANCE YES	NO	N/A ()	REMARKS () SEE BELOW
1. Is a procedure maintained that defines design practices and configuration management? (/207, 208b)	1						
2. Is a procedure maintained to provide for the evaluation of the engineering adequacy of design drawings? (4.1/207)	1						
3. Is a procedure maintained to provide for the evaluation of the engineering adequacy of design specifications? (4.1/207)	1						
4. Does the evaluation encompass the adequacy in relation to standard engineering and design practices? (4.1/207)	1						
5. Does the evaluation encompass the adequacy with respect to the design and purpose of the product? (4.1/207)	1						
6. Do procedures provide for identifying the point of effectivity for drawing changes? (4.1/208a)	1						
7. Is some means provided for recording the points of effectivity for changes employed? (4.1/208a)	1						
8. Is the record of drawing and documentation changes signed by an authorized person and maintained on file? (/208b)	3						
9. Are changes to design drawings evaluated for adequacy? (4.1/207)	1						
10. Are changes to design specifications evaluated for adequacy? (4.1/207)	1						
11. Are obsolete drawings and changes thereto removed from all points of issue and use? (4.1/208a)	3						
12. Is a procedure maintained that concerns itself with the currentness of drawings? (4.1/208a)	1						
13. Is the adequacy, currentness, and completeness of the following supplemental items reviewed to assure product conformity? (4.1/207)							
a. Specification	1						
b. Process Instructions	1						
c. Production Engineering Instructions	1						
d. Industrial Engineering Instructions	1						
e. Work Instructions	1						

B. DRAWINGS, DOCUMENTATION, AND CHANGES (cont.)	CODE	PROCEDURE/ DOCUMENT REVIEWED	NO. OF OBS MADE	COMPLIANCE		N/A ()	REMARKS () SEE BELOW
				YES	NO		
14. Does the program assure complete compliance with contract requirements for proposing, approving, and effecting engineering changes? (4.1/201)	1						
15. Does the program provide for monitoring compliance with approved engineering changes? (4.1/203)	1						
16. Does the program provide for submitting correct drawings and change information to the government in connection with data acquisition? (4.1)	1						
NOTE: When other specifications for waivers, deviations, configuration control, etc., are contractually required, i.e., MIL-STD-480/481/483, the contractor's implementation of these documents must also be reviewed.							

C. CALIBRATION (4.2, 4.3/209)	CODE	PROCEDURE/ DOCUMENT REVIEWED	NO. OF OBS MADE	COMPLIANCE		N/A ()	REMARKS () SEE BELOW
				YES	NO		
NOTE: The references in this section are to MIL-STD-45662A, Notice 3, and AQAP-6, Edition 2 respectively (XX/YY).	1						
1. Is there a written calibration system that provides adequate accuracy in the use of measuring and test equipment? (5.2/201, 205)	1						
2. Does the calibration system provide for the prevention of inaccuracy by detection of deficiencies and timely positive action for their correction? (5.1/205)	3						
3. Does the calibration system: (5.1/205)	3						
a. Include a listing of reference and transfer standards and their classification?							
b. Provide for nomenclature, identification number, calibration intervals, and source of calibration for reference and transfer standards?	3						
c. Include environmental conditions under which the measurement of standards will be applied and calibrated?	3						
d. Identify measuring and test equipment (M and TE) calibration intervals and sources?	3						
4. Do standards used for the calibration of measuring and test equipment have the capabilities for accuracy, stability, range, and resolution for the intended use? (5.2/203)	3						
5. Is the environment controlled to the extent necessary? (5.3/215)	4						
6. Do devices used to monitor environmental conditions display evidence of calibration? (/215)	3						
7. For the degree of accuracy required in the operation is it necessary to provide allowance for calibrations in other standard environments? (5.3/215)	4						
8. Is good housekeeping and cleanliness practiced in the calibration environment? (5.3/215)	4						
9. Are M and TE calibrated at intervals based on stability, purpose, degree of usage? (5.4/209)	1						
10. Are intervals adjusted based on results of previous calibrations? (5.4/209)	3						

C. CALIBRATION (4.2, 4.3/209) (cont.)

Item	CODE	PROCEDURE/ DOCUMENT REVIEWED	NO. OF OBS MADE	COMPLIANCE		N/A ()	REMARKS () SEE BELOW
				YES	NO		
11. Has a mandatory recall system been established for M and TE? (5.4/209)	1						
12. Are written procedures prepared and provided for calibration of all M and TE? (5.5/205)	1						
13. Do the instructions specify either the measurement standard to be used or the required accuracy of the standard and the accuracy of the instrument being calibrated? (5.5/206)	2						
14. Is the total calibration error (standard, equipment, personnel, procedure, environment) considered for determining the calibration accuracy? (/204, 214)	2						
15. Do the instructions require that calibration be performed by comparison with higher accuracy level standards? (5.5/205, 214)	2						
16. Are instructions being utilized? (5.5/205)	4						
17. Are adjustable devices on standards and measuring equipment that are fixed at the time of calibration sealed or otherwise safeguarded to prevent tampering by unauthorized personnel? (/208)	3						
18. Are procedures provided for using out-of-tolerance data to: (5.6.1/209)	1						
a. Adjust calibration frequency?							
b. Determine adequacy of M and TE?	1						
c. Determine adequacy of M and TE calibration procedures?	1						
19. Is M and TE which does not perform satisfactorily identified and its use prevented? (5.6.1/210)	1						
20. Are reporting channels identified for notification of significant out-of-tolerance conditions? (5.6.2/210)	1						
21. Do procedures define what constitutes a significant out-of-tolerance condition? (5.6.2/)	1						
22. Are standards used for calibration of M and TE certified as being traceable to the National or International Bureau of Standards? (5.7.1, 5.7.2/213)	3						
23. Are standards supported by certificates, reports, or data sheets attesting to date, accuracy, and environmental condition? (5.7.1/213)	3						

C. CALIBRATION (4.2, 4.3/209) (cont.)

Question	CODE	PROCEDURE/ DOCUMENT REVIEWED	NO. OF OBS MADE	COMPLIANCE YES	COMPLIANCE NO	N/A ()	REMARKS () SEE BELOW
24. If subcontractors are used, does the quality program assure that sources other than government or National Standards labs are capable of performing the service to the satisfaction of:	3						
a. MIL-STD-45662? (5.7.1, 5.10)							
b. AQAP-6? (/211)	3						
25. Are the applicable requirements of MIL-STD-45662/AQAP-6 delegated through purchase agreements when calibration is performed by subcontractors? (5.2/211)	3						
26. Are records maintained to assure accuracy of all M and TE? (5.8/209)	3						
27. Do the records of calibration include as a minimum: (5.8/206)	3						
a. Description or identification of the item?	3						
b. Calibration interval?	3						
c. Date of last calibration?	3						
d. Calibration results of out-of-tolerance condition?	3						
28. If the M and TE or measurement standard accuracy must be reported via a calibration report or certificate, does the record cite the report or certificate number or a copy on file? (5.8/206)	3						
29. Are labels or other means provided to monitor equipment adherence to calibration schedules? (5.9/207)	3						
30. Does the system indicate date of last calibration, by whom calibrated, and when the next calibration is due? (5.9/206)	3						
31. Are items which are not calibrated to their full capability or which require functional check only labeled to indicate the applicable condition? (5.9/206)	3						
32. Is M and TE stored, handled, and transported in a manner which shall not adversely affect the calibration or condition of the equipment? (5.11/212)	2						
33. Do procedures require the maintenance of gages and other M and TE necessary to assure product conformance to purchase order/contract agreements? (MIL-Q-9858A, para 4.2/212)	1						

C. CALIBRATION (4.2, 4.3/209) (cont.)	CODE	PROCEDURE/ DOCUMENT REVIEWED	NO. OF OBS MADE	COMPLIANCE		N/A (/)	REMARKS (/) SEE BELOW
				YES	NO		
34. Do procedures require that if production tooling is used as a media of inspection, that such devices are proven for accuracy at established intervals? (MIL-Q-9858A, para 4.3/201)	1						
35. If the calibration requirement exceeds the known state of the art is the customer/government notified? (MIL-Q-9858A, para 4.5/203)	2						
36. Is the calibration system periodically and systematically reviewed to ensure its effectiveness? (/202)	2						

D. SAMPLING INSPECTION

D. SAMPLING INSPECTION	CODE	PROCEDURE/ DOCUMENT REVIEWED	NO. OF OBS MADE	COMPLIANCE		N/A (/)	REMARKS (/) SEE BELOW
				YES	NO		
1. Do statistical methods and sampling procedures used conform with specification and contract requirements? (6.6/214)	1						
NOTE: Statistical sampling on contracts that are directly with the government must be in accordance with MIL-STD-105 or MIL-STD-414 unless otherwise approved by the government agency.							
2. Is the use of additional statistical methods other than those required by the contract suitable to maintain the required control of quality? (6.6/214)	2						
3. Has the customer approved the use of sampling plans that are not in accordance with contract requirements? (/214)	4						

E. PURCHASING	CODE	PROCEDURE/ DOCUMENT REVIEWED	NO. OF OBS MADE	COMPLIANCE		N/A ()	REMARKS () SEE BELOW
				YES	NO		
1. Are work instructions affecting quality prescribed in clear and complete documented work instructions? (3.3/201)	1						
2. Do the work instructions provide criteria for performing the work functions? (3.3/201)	3						
3. Does the program assure that all supplies and services procured from subcontractors conform to contract requirements? (5.1/210c)	1						
4. Does the program provide for the selection of subcontractors dependent on their demonstrated ability to perform and the quality evidence made available? (5.1/210a)	1						
5. Is objective evidence of quality furnished by the subcontractors used to the fullest extent? (5.1/210c)	2						
6. Is the effectiveness and integrity of the control of quality by subcontractors assessed and reviewed? (5.1/210a)	1						
7. Is this assessment and review accomplished at intervals consistent with the complexity and quantity of product? (5.1/210a)	2						
8. Is objective evidence of quality used in assessing subcontractor's performance? (5.1/210a)	2						
9. Are procedures established for the selection of qualified subcontractors? (5.1/201, 210a)	1						
10. Do the procedures provide for the transmission of design requirements to subcontractors? (5.1/210b)	1						
11. Do the procedures provide for the transmission of quality requirements to subcontractors? (5.1/210b)	1						
12. Does the quality program provide for the evaluation of the adequacy of procured items? (5.1/210c)	1						
13. Are the procedures adequate for providing subcontractors with feedback information on nonconforming supplies? (5.1/210c)	1						
14. Are the procedures adequate for assuring that subcontractors provide correction of nonconformances? (5.1/210c)	1						

E. PURCHASING (cont.)

	CODE	PROCEDURE/ DOCUMENT REVIEWED	NO. OF OBS MADE	COMPLIANCE		N/A ()	REMARKS () SEE BELOW
				YES	NO		
15. When certification of any type is required or utilized as a part of the basis for acceptance of procured materials, supplies, or processed items, are such documents reviewed for compliance to requirements, identified to the order in question, and filed? (3.4, 5.1/210c)	2						
16. Do purchase orders to subcontractors contain the following: (5.2/208a, 210b)	3						
a. A complete description of supplies ordered?							
b. Requirements for:	3						
1. Manufacturing?							
2. Inspection?							
3. Testing?							
4. Qualifications or approvals?							
5. Government or contractor inspection at source?							
6. Packaging?							
7. Certification?							
c. All pertinent technical requirements?	3						
17. Are subcontractors required to exercise control of raw materials to the extent necessary to assure conformance to technical requirements? (5.2/210a)	2						
18. If a Government Representative requires Government Inspection at a supplier, does the purchasing document to that supplier contain the following statement? (7.1/)	2						
"Government inspection is required prior to shipment from your plant. Upon receipt of this order, promptly notify the Government Representative who normally services your plant so that appropriate planning for Government inspection can be accomplished."							

E. PURCHASING (cont.)	CODE	PROCEDURE/ DOCUMENT REVIEWED	NO. OF OBS MADE	COMPLIANCE		N/A ()	REMARKS () SEE BELOW
				YES	NO		
19. When the statement in Question 18 (above) is authorized does that same purchasing document contain a statement substantially as follows? (7.1/) "On receipt of this order, promptly furnish a copy to the Government Representative who normally services your plant, or, if none, to the nearest Army, Navy, Air Force, or Defense Supply Agency inspection office. In the event the representative or office cannot be located, our purchasing agent should be notified immediately."	2						

F. RECEIVING INSPECTION	CODE	PROCEDURE/ DOCUMENT REVIEWED	NO. OF OBS MADE	COMPLIANCE		N/A ()	REMARKS () SEE BELOW
				YES	NO		
1. Are subcontracted supplies and materials subjected to inspection upon receipt to the extent necessary to assure conformance to technical requirements? (6.1/210c)	1						
2. Is the extent of the receiving effort adjusted based on objective quality data? (6.1/210c)	1						
3. Does the quality program provide for assurance that raw materials conform to technical requirements? (6.1/210c, 212)	1						
4. Are raw materials awaiting testing identified or segregated? (6.1/210c)	1						
5. Are controls established to prevent the inadvertent use of material failing to pass tests? (6.1/210c)	1						
6. Do records in receiving inspection indicate the: (3.4/205d)	3						
a. Part number? (/210d)	3						
b. Purchase order number or supplier number? (5.1/210a)	3						
c. Inspector identification? (/216)	3						
d. Nature of the observation made? (3.4/210d)	3						
e. Number of observations made? (3.4/210d)	3						
f. Number of deficiencies found? (3.4/210d)	3						
g. Type of deficiencies found? (3.4/210d)	3						
h. Quantities approved? (3.4/210d)	3						
i. Quantities rejected? (3.4/210d)	3						
j. Date? (5.1/210a)	3						
k. Nature of corrective action taken, as appropriate? (3.4/210d)	3						
7. Is a positive system maintained for identifying the inspection status of supplies? (6.7/216)	1						

G. STORES	CODE	PROCEDURE/ DOCUMENT REVIEWED	NO. OF OBS MADE	COMPLIANCE		N/A (√)	REMARKS (√) SEE BELOW
				YES	NO		
1. Are adequate work instructions provided to protect the quality of products during storage? (6.4/201, 217b)	1						
2. Are products subject to deterioration or corrosion cleaned and preserved to prevent these conditions? (6.4/217b)	1						
3. Are stored items inspected to detect deterioration or damage? (6.4/217b)	1						
4. Is the inspection of stored items documented? (3.4, 6.4)	2						
5. Are materials tested and approved kept identified until such time as its identity is necessarily obliterated by processing? (6.1/217)	3						
6. Is a positive system maintained for identifying the inspection status of stored supplies? (6.7/216, 217b)	1						

H. FABRICATING/PROCESSING	CODE	PROCEDURE/ DOCUMENT REVIEWED	NO. OF OBS MADE	COMPLIANCE YES	COMPLIANCE NO	N/A ()	REMARKS () SEE BELOW
1. Are work instructions affecting quality clear and complete and of a type appropriate to the circumstances for:	1						
a. Handling? (3.3/217a)	1						
b. Machining? (3.3/211a)	1						
c. Assembly? (3.3)	1						
d. Fabrication? (3.3/211a)	1						
e. Processing? (3.3/211b)	1						
f. Inspection? (3.3/211, 213a, 205c)	1						
2. Are production processes accomplished under controlled conditions including: (6.2/211b)	3						
a. Documented work instructions?	3						
b. Adequate production equipment?	3						
c. Special working environments?	3						
3. Is production fabrication accomplished under controlled conditions including: (6.2/211a)	3						
a. Documented work instructions?	3						
b. Adequate production equipment?	3						
c. Special working environments?	3						
4. Do work instructions include the criteria for acceptable or unacceptable workmanship in: (6.2/211a)	2						
a. Production processing?	2						
b. Production fabrication?	2						
5. Does the quality program provide for effective monitoring of the issuance of work instructions? (6.2/201, 211a)	1						
6. Does the quality program provide for the effective monitoring of the compliance with these work instructions? (6.2/201)	1						
7. When physical inspection of processed material is impossible or disadvantageous, is indirect control by monitoring equipment and personnel provided? (6.2/211a)	2						

H. FABRICATING/PROCESSING (cont.)	CODE	PROCEDURE/ DOCUMENT REVIEWED	NO. OF OBS MADE	COMPLIANCE		N/A (/)	REMARKS (/) SEE BELOW
				YES	NO		
8. Are both physical inspection and process monitoring used when required by contract or specification? (6.2/211a)	2						
9. Are methods of inspection and monitoring corrected when their unsuitability with reasonable evidence is demonstrated? (6.2/211a)	2						
10. Is adherence to selected methods of inspection and monitoring complete and continuous? (6.2/211a)	2						
11. Is a positive system maintained for identifying the inspection status of supplies during processing and fabrication? (6.7/216)	1						
12. Do records of in-process inspection indicate the: (3.4/205d)	3						
a. Part number? (/210d)							
b. Inspector identification? (/216)	3						
c. Nature of the observation made? (3.4/210d)	3						
d. Number of observations made? (3.4/210d)	3						
e. Number of deficiencies found? (3.4/210d)	3						
f. Type of deficiencies found? (3.4/210d)	3						
g. Quantities approved? (3.4/210d)	3						
h. Quantities rejected? (3.4/210d)	3						
i. Date? (6.2/)	3						
j. Nature of corrective action taken, as appropriate? (3.4/210d)	3						
13. Does the quality program require the use of procedures to instruct individuals how to prevent handling damage to articles? (6.4/217a)	1						

I. FINAL INSPECTION	CODE	PROCEDURE/ DOCUMENT REVIEWED	NO. OF OBS MADE	COMPLIANCE		N/A (/)	REMARKS () SEE BELOW
				YES	NO		
1. Are work instructions affecting quality clear and complete and of a type appropriate to: (3.3/213b, 205c)	1						
a. Final testing?	1						
b. Final inspection?	1						
2. Does the quality program assure that there is a system for final inspection and test of completed products? (6.3/213b)	1						
3. Does such testing provide a measure of overall product quality? (6.3/213b)	1						
4. Is final testing performed so that it simulates to a sufficient degree, product end use and functioning? (6.3)	1						
5. When unusual difficulties, deficiencies, or questionable conditions are found during final inspection and testing, are they reported to:	1						
a. Designers? (6.3/207)							
b. The customer? (/213c)	1						
6. When modifications, repairs, or replacements are required after final inspection or testing, is there reinspection and retesting of any characteristics affected? (6.3)	2						
7. Do records of final inspection indicate the: (3.4/205d)	3						
a. Part number? (/210d)	3						
b. Inspector identification? (/216)	3						
c. Nature of the observation made? (3.4/210d)	3						
d. Number of observations made? (3.4/210d)	3						
e. Number of deficiencies found? (3.4/210d)	3						
f. Type of deficiencies found? (3.4/210d)	3						
g. Quantities approved? (3.4/210d)	3						
h. Quantities rejected? (3.4/210d)	3						
i. Date? (6.3/)	3						
j. Nature of corrective action taken, as appropriate? (3.4/210d)	3						

J. SHIPPING	CODE	PROCEDURE/ DOCUMENT REVIEWED	NO. OF OBS MADE	COMPLIANCE		N/A (√)	REMARKS (√) SEE BELOW
				YES	NO		
1. Are adequate work instructions provided to protect the quality of products during: (6.4/217c)							
a. Preservation?	1						
b. Packaging?	1						
c. Shipping?	1						
2. Are adequate inspection instructions provided to assure the quality of products: (6.4/217c)	1						
a. Preserved?	1						
b. Packaged?	1						
c. Shipped?	1						
3. Are products subject to deterioration or corrosion cleaned and preserved to prevent these conditions? (6.4/217)	1						
4. Does the quality program provide for monitoring of shipments to assure that products shipped are in compliance with contract requirements and other shipping regulations? (6.4)	1						
5. Is a positive system maintained for identifying the inspection status of supplies during preservation, packaging, and shipping? (6.7/216)	1						
6. Do records of preservation, packaging, packing, and marking indicate the: (3.4/205d)	3						
a. Part number? (/210d)	3						
b. Contract number? (6.4/217a)	3						
c. Inspector identification? (/216)	3						
d. Quantities? (3.4/210d)	3						
e. Date? (6.4/217c)	3						

K. NONCONFORMING MATERIAL

	CODE	PROCEDURE/ DOCUMENT REVIEWED	NO. OF OBS MADE	COMPLIANCE		N/A (∅)	REMARKS (∅) SEE BELOW
				YES	NO		
1. Are procedures acceptable to the government/customer that provide for the (a/b/c) of nonconforming material? (6.5/215)	1						
a. Identification							
b. Segregation	1						
c. Disposition	1						
2. Is repair or rework of nonconforming supplies accomplished by documented procedures acceptable to the government/ customer? (6.5/215)	1						
3. Are holding areas or procedures provided and acceptable to the government/customer? (6.5/215)	1						

L. CORRECTIVE ACTION	CODE	PROCEDURE/ DOCUMENT REVIEWED	NO. OF OBS MADE	COMPLIANCE		N/A ()	REMARKS () SEE BELOW
				YES	NO		
1. Does the quality program provide for prompt detection of defective supplies or other elements of contract performance? (3.5/206)	1						
2. Is corrective action taken to correct assignable conditions that have resulted in defective supplies or other elements of contract performance? (3.5/206)	1						
3. Is corrective action taken to correct assignable conditions that could result in defective supplies or other elements of contract performance? (3.5/206)	1						
4. Are corrective action requirements extended to the performance of all suppliers? (3.5/206)	1						
5. Is the corrective action program responsive to data and product deficiencies received from users? (3.5/206)	1						
6. Is correct action taken as a result of product and data analysis of scrapped or reworked supplies? (3.5/206)	2						
7. Is corrective action taken as a result of the analysis of trends in processes to prevent nonconforming products? (3.5/)	1						
8. Is corrective action taken as a result of the analysis of trends in the performance of work to prevent nonconforming products? (3.5/)	1						
9. When corrective measures have been taken, does the contractor review and monitor these measures for adequacy and effectiveness? (3.1, 3.5/206)	2						

M. COST OF QUALITY	CODE	PROCEDURE/ DOCUMENT REVIEWED	NO. OF OBS MADE	COMPLIANCE		N/A (/)	REMARKS (/) SEE BELOW
				YES	NO		
1. Is cost data identifying cost of preventing nonconforming supplies maintained? (3.6)	1						
2. Is cost data identifying cost of correcting nonconforming supplies maintained? (3.6)	1						
3. Is the cost data maintained and identified to the above items? (3.6)	2						
4. Is the cost data generated used as a management element of the quality program? (3.6)	2						
5. Are scrap and rework cost and loss data available? (6.5)	4						
6. Is the cost data available, upon request, to the government representative for "on-site" review? (3.6)	4						

N. GOVERNMENT PROPERTY	CODE	PROCEDURE/ DOCUMENT REVIEWED	NO. OF OBS MADE	COMPLIANCE		N/A (/)	REMARKS (/) SEE BELOW
				YES	NO		
1. If the supplier has Government Furnished Material (GFM) are there procedures that include provisions for: (7.2.1/)	1						
a. Examination on receipt for damage in transit?	1						
b. Inspection for completeness and proper type?	1						
c. Periodic inspection of stored GFM to determine deterioration or damage?	1						
d. Functional testing either prior to or after installation to determine satisfactory operation?	1						
e. Identification and protection from improper use or disposition?	1						
2. Does the supplier report to the Government Representative any GFM found damaged, malfunctioning, or otherwise unsuitable for use? (7.2.2/)	3						
3. Does the supplier determine and record the probable cause for damaged or malfunctioning GFM? (7.2.2/)	2						
4. Does the supplier determine and record the necessity for withholding from use damaged or malfunctioning GFM? (7.2.2/)	2						
5. Are procedures established to assure adequate storage of Bailed Property? (7.2.3/)	1						
6. Are procedures established to assure adequate maintenance of Bailed Property? (7.2.3/)	1						
7. Are procedures provided to assure adequate inspection of Bailed Property? (7.2.3/)	1						
8. Are inspection and maintenance records of Bailed Property maintained? (7.2.3/)	2						

APPENDIX B

MIL-I-45208A/AQAP-4 AUDIT WORKSHEET

Worksheet Instructions

The following information will clarify the use of this worksheet:

Codes

1 — Identify the written procedure. A procedure is a document that specifies or describes how an activity is to be performed.

2 — Identify the instruction. An instruction is an order or direction outlining a sequence of steps to follow in performing a specific inspection, test, or work function.

3 — Verify documentation. Documentation is anything printed or written that is relied upon to prove or support, e.g., purchase order, forms, flow charts, records, travelers, drawings, memos, etc. A written explanation for a uniform application and preparation of documentation may be required.

4 — A minimum of three observations shall be performed.

Procedure/Document Reviewed

When Codes 1 or 2 are identified the auditee is to record the applicable procedure/instruction in this column. This is to be done before the survey/audit.

Number of Observations Made

When Code 4 is identified the auditor is to record the actual number of observations made in this column.

Compliance — Yes/No

After evaluation of each applicable question the auditor is to identify whether or not the auditee is in compliance.

N/A (✓)

This column is to be checked (✓) if the question is not applicable to the auditee.

Remarks (✓) See Below

This column is to be checked (✓) if further explanation is necessary. The explanation is to be given at the end of the applicable section and identified to the question number.

Specification/Checklist Cross-Reference

MIL-I-45208A		AQAP-4	
3.1	A1, A2, A7, C3, D1, D2, J2	201	A1, A6, J1
3.2.1	A8, A9, E6, E7, G3, H1	202	A3
3.2.2	A10, A11, E8, E10, G4, H2, I5	203a	A8, A9, A13, E6, E7, G6, G7, G9, H1, H3, I1
3.2.3	K1–7		
3.2.4	A12, A13, E9, G5–7, H3	203b	A10, A11, E8, E10, G4, H2, I5
3.3	-B-		
3.4	G9	203c	A12, E9, G5, G6, H3
3.5	A14–16, F3, G10, H4, I4	204	-B-
3.6	A17, A18, E14, F4	205a	D2, D3, E3, E11, E12
3.6.1	A19	205b	D1
3.7	J1, J3, J4	205c	E1
3.8	E2	205d	D7
3.9	C1, C2	206	A7, G5, G6
3.10	A20	207	G5, G8
3.11	A4, A21	208a	A14, A15, E13, F3, G10, H4, I4
3.11.1	D4		
3.11.2	D5	208b	E13, F1, F4
3.11.3	A22	208c	E4, E5, F2, G1, G2, I2
3.12	D6, E1	209	C1, C2
3.13	A3	210	J1, J3, J4
		211	A7
		212	I1, I2
		213	K1–7
		214	A20
		215	A2–4, A22

Specification/Checklist Cross-Reference

Survey Category "B"
(Calibration)

MIL-STD-45662A	AQAP-6
5.1 B2, B3	201 B1, B34
5.2 B1, B4, B25	202 B36
5.3 B5, B7, B8	203 B4, B35
5.4 B9–11	204 B14, B33
5.5 B12, B13, B15, B16	205 B1–3, B12, B15, B16
5.6.1 B18, B19	206 B13, B27, B28, B30, B31
5.6.2 B20, B21	207 B29
5.7.1 B22–24	208 B17
5.7.2 B22	209 B9–11, B18, B26
5.8 B26–28	210 B19, B20
5.9 B29–31	211 B24, B25
5.10 B24	212 B32
5.11 B32	213 B22, B23
	214 B14, B15
	215 BC5–8
	216
	217

A. QUALITY SYSTEM	CODE	PROCEDURE/ DOCUMENT REVIEWED	NO. OF OBS MADE	COMPLIANCE YES	COMPLIANCE NO	N/A ()	REMARKS () SEE BELOW
1. Is the supplier's inspection system documented? (3.1/201)	1						
2. Is all documentation made available for review? (3.1/215)	4						
3. Does the supplier permit evaluation of their quality system and the products/services produced? (3.13/202, 215)	4						
4. When on-site verification of purchase order conformance is required, does the supplier provide the equipment facilities and personnel necessary for verification? (3.11/215)	4						
5. Does the supplier maintain a single quality system?	4						
6. Has the responsibility for quality control/inspection been formally established? (/201)	1						
7. Do the procedures require inspection to substantiate product/ service conformance? (3.1/206, 211)	1						
8. Do procedures require inspections to be performed by clear and complete instructions, as required by specification and contract? (3.2.1/203a)	1						
9. Do procedures require testing to be performed by clear and complete instructions, as required by specification and contract? (3.2.1/203a)	1						
10. Do procedures require the supplier to maintain inspection records? (3.2.1/203a)	1						
11. Are inspection records retained on file as required by contract/ purchase order? (3.2.2/203b)	3						
12. Do procedures assure that the latest applicable drawings and authorized changes are used? (3.2.4/203c)	1						
13. Do procedures assure that the latest applicable specifications and authorized changes are used? (3.2.4/203a)	1						
14. Do procedures require a positive system for identifying the inspection status (accepted, rejected, pending) of products? (3.5/208a)	1						
15. If inspection stamps are used, is an adequate issue record maintained? (3.5/208a)	3						
16. If inspection stamps are used, are they uniquely different from customer/government stamps? (3.5/)	4						

A. QUALITY SYSTEM (cont.)	CODE	PROCEDURE/ DOCUMENT REVIEWED	NO. OF OBS MADE	COMPLIANCE		N/A (/)	REMARKS (/) SEE BELOW
				YES	NO		
17. Do procedures require functional testing of customer/government material (FM) prior to or after installation to determine satisfactory operation? (3.6/)	1						
18. Do procedures require identification, segregation, and protection of FM from improper use or disposition? (3.6/)	1						
19. Do procedures require the supplier to report to the customer and FM found damaged, malfunctioning, or otherwise unsuitable for use? (3.6.1/)	1						
20. Are alternative inspection instructions and/or equipment utilized and, if so, are they adequate? (3.10/214)	1						
21. Is government inspection at subcontractor's facilities only requested by or authorized by government representative? (3.11/)	3						
22. Are all purchase documents and referenced data made available to the government representative? (3.11.2, 3.11.3./215)	4						

B. CALIBRATION (3.3/204)	CODE	PROCEDURE/ DOCUMENT REVIEWED	NO. OF OBS MADE	COMPLIANCE		N/A (∅)	REMARKS (∅ SEE BELOW)
				YES	NO		
NOTE: The references in this section are to MIL-STD-45662A, Notice 3 and AQAP-6, Edition 2 respectively (XX/YY).							
1. Is there a written calibration system that provides adequate accuracy in the use of measuring and test equipment? (5.2/201, 205)	1						
2. Does the calibration system provide for the prevention of inaccuracy by detection of deficiencies and timely positive action for their correction? (5.1/205)	1						
3. Does the calibration system: (5.1/205)	3						
a. Include a listing of reference and transfer standards and their classification?							
b. Provide for nomenclature, identification number, calibration intervals and source of calibration for reference and transfer standards?	3						
c. Include environmental conditions under which the measurement of standards will be applied and calibrated?	3						
d. Identify measuring and test equipment (M and TE) calibration intervals and sources?	3						
4. Do standards used for the calibration of measuring and test equipment have the capabilities for accuracy, stability, range, and resolution for the intended use? (5.2/203)	3						
5. Is the environment controlled to the extent necessary? (5.3/215)	4						
6. Do devices used to monitor environmental conditions display evidence of calibration? (/215)	3						
7. For the degree of accuracy required in the operation is it necessary to provide allowance for calibrations in other standard environments? (5.3/215)	4						
8. Is good housekeeping and cleanliness practiced in the calibration environment? (5.3/215)	4						
9. Are M and TE calibrated at intervals based on stability, purpose, degree of usage? (5.4/209)	1						
10. Are intervals adjusted based on results of previous calibrations? (5.4/209)	3						

B CALIBRATION (3.3/204) (cont.)

	CODE	PROCEDURE/ DOCUMENT REVIEWED	NO. OF OBS MADE	COMPLIANCE YES	COMPLIANCE NO	N/A ()	REMARKS () SEE BELOW
11. Has a mandatory recall system been established for M and TE? (5.4/209)	1						
12. Are written procedures prepared and provided for calibration of all M and TE? (5.5/205)	1						
13. Do the instructions specify either the measurement standard to be used or the required accuracy of the standard and the accuracy of the instrument being calibrated? (5.5/206)	2						
14. Is the total calibration error (standard, equipment, personnel, procedure, environment) considered for determining the calibration accuracy? (/204,214)	2						
15. Do the instructions require that calibration be performed by comparison with higher accuracy level standards? (5.5/205, 214)	2						
16. Are instructions being utilized? (5.5/205)	4						
17. Are adjustable devices on standards and measuring equipment that are fixed at the time of calibration sealed or otherwise safe-guarded to prevent tampering by unauthorized personnel? (/208)	3						
18. Are procedures provided for using out-of-tolerance data to: (5.6.1/209)	1						
a. Adjust calibration frequency?							
b. Determine adequacy of M and TE?	1						
c. Determine adequacy of M and TE calibration procedures?	1						
19. Is M and TE, which does not perform satisfactorily, identified and its use prevented? (5.6.1/210)	1						
20. Are reporting channels identified for notification of significant out of tolerance conditions? (5.6.2/210)	1						
21. Do procedures define what constitutes a significant out of tolerance condition? (5.6.2/)	1						
22. Are standards used for calibration of M and TE certified as being traceable to the National or International Bureau of Standards? (5.7.1, 5.7.2/213)	3						
23. Are standards supported by certificates, reports, or data sheets attesting to date, accuracy, and environmental condition? (5.7.1/213)	3						

B. CALIBRATION (3.3/204) (cont.)	CODE	PROCEDURE/ DOCUMENT REVIEWED	NO. OF OBS MADE	COMPLIANCE YES	NO	N/A ()	REMARKS () SEE BELOW
24. If subcontractors are used, does the quality program assure that sources other than government or National Standards labs are capable of performing the service to the satisfaction of:							
a. MIL-STD-45662? (5.7.1, 5.10)	3						
b. AQAP-6? (/211)	3						
25. Are the applicable requirements of MIL-STD-45662/AQAP-6 delegated through purchase agreements when calibration is performed by subcontractors? (5.2/211)	3						
26. Are records maintained to assure accuracy of all M and TE? (5.8/209)	3						
27. Do the records of calibration include as a minimum: (5.8/206)	3						
a. Description or identification of the item?							
b. Calibration intervals?	3						
c. Date of last calibration?	3						
d. Calibration results of out-of-tolerance condition?	3						
28. If the M and TE or measurement standard accuracy must be reported via a calibration report or certificate, does the record cite the report or certificate number or a copy on file? (5.8/206)	3						
29. Are labels or other means provided to monitor equipment adherence to calibration schedules? (5.9/207)	3						
30. Does the system indicate date of last calibration, by whom calibrated, and when the next calibration is due? (5.9/206)	3						
31. Are items which are not calibrated to their full capability or which require functional check only labeled to indicate the applicable condition? (5.9/206)	3						
32. Is M and TE stored, handled, and transported in a manner which shall not adversely affect the calibration or conditions of the equipment? (5.11/212)	2						
33. Do procedures require the maintenance of gages and other M and TE necessary to assure product conformance to purchase order/contract agreements? (MIL-I-45208A, para 3.3/204)	1						

B. CALIBRATION (3.3/204) (cont.)	CODE	PROCEDURE/ DOCUMENT REVIEWED	NO. OF OBS MADE	COMPLIANCE		N/A ()	REMARKS () SEE BELOW
				YES	NO		
34. Do procedures require that if production tooling is used as a media of inspection, that such devices are proven for accuracy at established intervals? (/201)	1						
35. If the calibration requirement exceeds the known state of the art is the customer/government notified? (/203)	2						
36. Is the calibration system periodically and systematically reviewed to ensure its effectiveness? (/202)	2						

C. SAMPLING INSPECTION	CODE	PROCEDURE/ DOCUMENT REVIEWED	NO. OF OBS MADE	COMPLIANCE		N/A (/)	REMARKS (/) SEE BELOW
				YES	NO		
1. Do statistical methods and sampling procedures used conform with specification and contract requirements? (3.9/209)	1						
2. If not stated in the contract, has the sampling inspection been subject to approval by the government? (3.9/209)	4						
3. Do inspection personnel have instructions covering sampling inspection? (3.1/)	2						
4. What sampling plan and AQLs are used?	4						

4. What sampling plan and AQLs are used?

MIL-STD-105 [ ] AQL [ ]

MIL-STD-414 [ ] AQL [ ]

DODGE-ROMIG [ ] AQL [ ]

OTHER [ ] AQL [ ]

D. PURCHASING	CODE	PROCEDURE/ DOCUMENT REVIEWED	NO. OF OBS MADE	COMPLIANCE		N/A ()	REMARKS () SEE BELOW
				YES	NO		
1. Do procurement documents reflect drawing and/or specification requirements? (3.1/205b)	3						
2. Are customer requirements/changes reflected on outgoing procurement documents? (3.1/205a)	3						
3. Are certified test reports or certificates of compliance required by procurement documents? (/205a)	3						
4. When government inspection is required, does the supplier add the appropriate statement to the purchasing documents? (3.11.1/)	3						
5. When the government representative authorizes copies of the purchasing document to be furnished to the government representative at the subcontractor's facility, is the appropriate statement added to the purchasing documents? (3.11.2/)	3						
6. Does the supplier report any nonconformances found in customer/ government source inspected supplies? (3.12/)	1						
7. Do purchasing documents reserve the right to verify at the supplier's facility that purchased material conforms with requirements? (/205d)	3						

E. RECEIVING INSPECTION	CODE	PROCEDURE/ DOCUMENT REVIEWED	NO. OF OBS MADE	COMPLIANCE		N/A (/)	REMARKS (/) SEE BELOW
				YES	NO		
1. Are subcontracted supplies and materials subjected to inspection upon receipt to the extent necessary to assure conformance to technical requirements? (3.12/205c)	1						
2. Are qualified products (QPL items) inspected and tested to assure conformance to specification requirements? (3.8/)	4						
3. When certifications of any type are required or utilized as a part of the basis for acceptance of procured materials, supplies, or processed items, are such documents reviewed for compliance to requirements, identified to the order in question, and filed? (/205a)	3						
4. Is good housekeeping maintained in the manufacturing and receiving inspection area? (/208c)	4						
5. Are products properly protected and handled to prevent inadvertent damage, contamination, and/or deterioration? (/208c)	4						
6. Are clear and complete instructions, as required by specification and contract, used to:	2						
a. Inspect parts? (3.2.1/203a)	2						
b. Test parts? (3.2.1/203a)	2						
c. Identify accept/reject criteria for parts? (3.2.1/203a)	2						
7. Are clear and complete instructions, as required by specification and contract, used to:	2						
a. Inspect raw material? (3.2.1/203a)	2						
b. Test raw materials? (3.2.1/203a)	2						
c. Identify accept/reject criteria for raw materials? (3.2.1/203a)	3						
8. Do records in receiving inspection indicate the:	3						
a. Part number? (/203b)	3						
b. Contract number or supplier number? (/203b)	3						
c. Nature of the observation made? (3.2.2/203b)	3						
d. Number of observations made? (3.2.2/203b)	3						
e. Number of deficiencies found? (3.2.2/203b)							

E. RECEIVING INSPECTION (cont.)	CODE	PROCEDURE/ DOCUMENT REVIEWED	NO. OF OBS MADE	COMPLIANCE		N/A ()	REMARKS () SEE BELOW
				YES	NO		
f. Type of deficiencies found? (3.2.2/203b)	3						
g. Quantities approved? (3.2.2/203b)	3						
h. Quantities rejected? (3.2.2/203b)	3						
i. Date?	3						
j. Nature of corrective action taken, as appropriate? (3.2.2/203b)	3						
9. Are the following documents available to the latest applicable revision or authorized change?	1-4						
a. Drawings (3.2.4/203c)							
b. Specifications (3.2.4/203c)	4						
c. Instructions/procedures (3.2.4/203c)	4						
10. Are materials identified to the applicable purchase order or material certification, heat or melt number when required? (3.2.2/203b)	3						
11. Do procedures require that material, when accepted on test reports and/or certificates of conformance, be subjected to verification testing? (/205a)	1						
12. Are test reports or certificates of chemical and physical analysis maintained on file? (/205a)	4						
13. Do procedures require that inspected items be properly segregated from material awaiting inspection? (/208b)	1-3						
a. Verify the inspection status is indicated on parts. (/208a)							
b. Verify the inspection status is indicated on raw material. (/208a)	3						
14. Do procedures include provisions for customer/government furnished material (FM) in the following areas:	1						
a. Examination on receipt for damage in transit? (3.6/)							
b. Inspection for completeness and proper type? (3.6/)	1						
c. Verification of quantity? (3.6/)	1						

F. STORES	CODE	PROCEDURE/ DOCUMENT REVIEWED	NO. OF OBS MADE	COMPLIANCE		N/A (/)	REMARKS (/) SEE BELOW
				YES	NO		
1. Is "good housekeeping" maintained? (/208b)	4						
2. Are adequate work instructions provided to protect the quality of products during storage? (/208c)	4						
3. Is a positive system maintained for identifying the inspection status of stored supplies? (3.5/208a)	1						
4. Do procedures require the supplier to include provisions for customer/government furnished material (FM) in the following areas?:	1						
a. Periodic inspection of stored FM to determine deterioration or damage. (3.6/)							
b. Precaution to assure adequate storage and handling conditions. (3.6/208b)	1						

G. FABRICATING/PROCESSING	CODE	PROCEDURE/ DOCUMENT REVIEWED	NO. OF OBS MADE	COMPLIANCE		N/A (/)	REMARKS (/) SEE BELOW
				YES	NO		
1. Is "good housekeeping" maintained? (/208c)	4						
2. Are products properly protected and handled to prevent inadvertent damage, contamination, and/or deterioration? (/208c)	4						
3. Are clear and complete instructions, as required by specification and contract, used to:	2						
a. Inspect parts? (3.2.1/)	2						
b. Test parts? (3.2.1/)	2						
c. Identify accept/reject criteria for parts? (3.2.1/)	2						
4. Do records of in-process inspection indicate the:	3						
a. Part number? (/203b)	3						
b. Nature of the observation made? (3.2.2/203b)	3						
c. Number of observations made? (3.2.2/203b)	3						
d. Number of deficiencies found? (3.2.2/203b)	3						
e. Type of deficiencies found? (3.2.2/203b)	3						
f. Quantities approved? (3.2.2/203b)	3						
g. Quantities rejected? (3.2.2/203b)	3						
h. Date?	3						
i. Nature of corrective action taken, as appropriate? (3.2.2/203b)	3						
5. Are the applicable drawings and authorized changes required by contract used during:	4						
a. In-process inspection? (3.2.4/203c, 206)	4						
b. Manufacturing/processing? (3.2.4/207)	4						
6. Are the applicable specifications and authorized changes required by contract used during:	4						
a. In-process inspection? (3.2.4/203c, 206)	4						
b. Manufacturing/processing? (/203a, 203c, 206)	4						

G. FABRICATING/PROCESSING (cont.)	CODE	PROCEDURE/ DOCUMENT REVIEWED	NO. OF OBS MADE	COMPLIANCE		N/A (/)	REMARKS (/) SEE BELOW
				YES	NO		
7. Instructions are current and available for:							
a. Inspection (3.2.4/203a)	2						
b. Manufacturing/processing (/203a)	2						
8. Has criteria for acceptable workmanship been established through written standards or representative samples? (/207)	2						
9. When process control procedures are required by contract or specification, are the procedures an integral part of the inspection system and do they conform to the specific contract/specification requirements? (3.4/203a)	1						
10. Verify that inspection status is indicated on parts during:	3						
a. Manufacturing (3.5/208a)							
b. Processing (3.5/208a)	3						

H. FINAL INSPECTION	CODE	PROCEDURE/ DOCUMENT REVIEWED	NO. OF OBS MADE	COMPLIANCE		N/A (/)	REMARKS (/) SEE BELOW
				YES	NO		
1. Are clear and complete instructions, as required by specification and contract, used to:	2						
a. Inspect parts? (3.2.1/203a)							
b. Test parts? (3.2.1/203a)	2						
c. Identify accept/reject criteria for parts? (3.2.1/203a)	2						
2. Do records of final inspection indicate the:	3						
a. Part number? (3.2.2/203b)							
b. Nature of the observation made? (3.2.2/203b)	3						
c. Number of observations made? (3.2.2/203b)	3						
d. Number of deficiencies found? (3.2.2/203b)	3						
e. Type of deficiencies found? (3.2.2/203b)	3						
f. Quantities approved? (3.2.2/203b)	3						
g. Quantities rejected? (3.2.2/203b)	3						
h. Date?	3						
i. Nature of corrective action taken, as appropriate? (3.2.2/203b)	3						
3. Are the following documents used to perform final inspection or test to the applicable revision or authorized change?	3						
a. Drawings (3.2.4/203a)							
b. Specifications (3.2.4/203c)	3						
c. Instructions/procedures (3.2.4/203c)	3						
4. Is the inspection status indicated on parts? (3.5/208a)	3						

I. SHIPPING	CODE	PROCEDURE/ DOCUMENT REVIEWED	NO. OF OBS MADE	COMPLIANCE		N/A ()	REMARKS () SEE BELOW
				YES	NO		
1. Are adequate inspection instructions provided to assure the quality of products:	2						
a. Preserved? (/203a, 212)	2						
b. Packaged? (/203a, 212)	2						
c. Shipped?	2						
2. Are products subject to deterioration or corrosion cleaned and preserved to prevent these conditions? (/208c, 212)	3						
3. Does the quality system provide for monitoring of shipments to assure that products shipped are in compliance with contract requirements and other shipping regulations? (/202)	1						
4. Is a positive system maintained for identifying the inspection status of supplies during preservation, packaging, and shipping? (3.5/208a)	1						
5. Do records of preservation, packaging, packing, and marking indicate the:	3						
a. Part number? (/203b)	3						
b. Contract number? (/203b)	3						
c. Nature of the observation made? (3.2.2/203b)	3						
d. Number of observations made? (3.2.2/203b)	3						
e. Number of deficiencies found? (3.2.2/203b)	3						
f. Type of deficiencies found? (3.2.2/203b)	3						
g. Quantities approved? (3.2.2/203b)	3						
h. Quantities rejected? (3.2.2/203b)	3						
i. Date?	3						
j. Nature of corrective action taken, as appropriate? (3.2.2/203b)	3						

J. NONCONFORMING MATERIAL	CODE	PROCEDURE/ DOCUMENT REVIEWED	NO. OF OBS MADE	COMPLIANCE		N/A (/)	REMARKS (/) SEE BELOW
				YES	NO		
1. Are procedures acceptable to the government/customer that provide for the (a/b/c/d) of nonconforming material?	1						
a. Identification (3.7/210)							
b. Segregation (3.7/210)	1						
c. Disposition (3.7/210)	1						
d. Presentation (3.1/201)	1						
2. Is repair or rework of nonconforming supplies accomplished by documented procedures acceptable to the government/customer? (3.1/)	3						
3. Are holding areas or procedures provided and acceptable to the government/customer? (3.7/210)	4						
4. Do procedures require control to prevent the use of nonconforming material? (3.7/210)	1						

K. CORRECTIVE ACTION	CODE	PROCEDURE/ DOCUMENT REVIEWED	NO. OF OBS MADE	COMPLIANCE		N/A ()	REMARKS () SEE BELOW
				YES	NO		
1. Is corrective action taken to correct assignable conditions that could result in defective supplies or other elements of contract performance? (3.2.3/213)	1						
2. Is corrective action taken to correct assignable conditions that have resulted in defective supplies or other elements of contract performance? (3.2.3/213)	1						
3. Are corrective action requirements extended to the performance of all suppliers? (3.2.3/213)	3						
4. Is the corrective action program responsive to data and product deficiencies received from users? (3.2.3/213)	2						
5. Is corrective action taken as a result of product and data analysis of scrapped or reworked supplies? (3.2.3/213)	2						
6. Is corrective action taken as a result of the analysis of trends in processes to prevent nonconforming products? (3.2.3/213)	2						
7. Is corrective action taken as a result of the analysis of trends in the performance of work to prevent nonconforming products? (3.2.3/213)	2						
8. When corrective measures have been taken, does the contractor review and monitor these measures for adequacy and effectiveness? (/202)	2						

REFERENCES

1. ANSI/ASQC A3-1987 *Quality Systems Terminology*. Milwaukee: American Society for Quality Control, 1987.

2. ISO 8402 *Quality — Vocabulary*. First edition 1986-06-15, 8402 ADD.1 1987-10-07. Paris: International Standards Organization.

3. Robinson, Charles B. *How to Plan an Audit*. Milwaukee: ASQC Quality Press, 1987.

4. Feigenbaum, Armand V. *Total Quality Control,* 3rd ed. New York: McGraw-Hill Book Co., 1983.

5. Chruden, H.J., and A.W. Sherman, Jr. *Personnel Management,* 5th ed. Cincinnati: South-Western Publishing Co., 1979.

6. National Standard of Canada CAN3-Q395-81. *Quality Audits*. Rexdale, Ontario: Canadian Standards Association, 1981.

7. Sayle, Allan J. *Management Audits*. New York: McGraw-Hill Book Co., 1985.

8. Thresh, James L. *How to Conduct, Manage, and Benefit from Effective Quality Audits*. Harrison, New York: MGI Management Institute, 1984.

9. Juran, Joseph M. *Quality Control Handbook,* 3rd ed. New York: McGraw-Hill Book Co., 1979.

10. NASI N45.2.10-1973 *Quality Assurance Terms and Definitions*. New York: American Society of Mechanical Engineers, 1973.

11. ANSI/ASQC Q1-1986 *Generic Guidelines for Auditing of Quality Systems*. Milwaukee: American Society for Quality Control, 1986.

12. *Generic Guidelines for Auditing Quality Systems*. Quality 1 Audit. Paris: International Organization for Standardization.

13. ANSI/ASQC Q90-1987 *Quality Management and Quality Assurance Standards — Guidelines for Selection and Use*. Milwaukee: American Society for Quality Control, 1987.

14. ASQC Programs Committee of the Energy Division. *Nuclear Quality Systems Auditor Training Handbook*. Milwaukee: American Society for Quality Control, 1980.

INDEX

Audit:
  Audit deficiency report, sample of, 37-38
  Audit exit interview report, sample of, 36
  Audit notification letter, sample of, 34
  Close-out audit, 67
  Corporate audits, 85-88
  Evaluation, x
  Internal audit, 43-44
  Policy, 31-32, 82
  Procedure, ix-x
  Product audits, 77-80
  Purpose of a quality system audit, 23-29
  Schedule, 32
  Special process audits, 81-83
  Supplier audit, 32-43
  Unified supplier base audits, 88
Audit report, 59-63
Audit worksheets, x, 45-57
  Addendum, 47-48
  Evaluation, 49,50
  MIL-I-45208A/AQAP-4 audit worksheet, 131-151
  MIL-Q-9858A/AQAP-1 audit worksheet, 105-130
  Scoring, 48-50
  Special process audits, 82-83
  Validation of, 49
Auditor:
  Independence, 70-71
  Performance, 57
  Qualifications, ix, 69-76, 83

Close-out audit, 67
Corporate audits, 85-88
Corrective action, 65-66

Defense specifications/standards, 1-2

Deficiency survey/audit deficiency reports, sample of, 37-38
Definitions used in auditing discipline, 93-104

Evaluation categories and rating, 51-56
Evaluation score, 48

Follow-up action, 66

Independence of auditor, 70-71
Inspection requirements, 3
Inspection system requirements for MIL-I-45208A, 8-10, 10-19
Internal audit, 43-44
Interview: survey/audit exit interview report, sample of, 36

MIL-I-45208A — Inspection system requirements, 8-10, 20
  Compared to MIL-Q-9858A, 10-19
MIL-I-45208A/AQAP audit worksheet, 131-151
MIL-Q-9858A — Quality program requirements, 5-8, 20
  Compared to MIL-I-45208A, 10-19
MIL-Q-9858A/AQAP audit worksheet, 105-130

Product audits, 77-80
  Frequency, 77-78
  Quality rating, 78-80
  Scope, 77

Quality assurance supplier status report, sample of, 35
Quality management, xii
Quality program requirements, 2
  Application of, 3-4
Quality requirements guide, 4

Quality system, defining, 1-4
Quality system audit:
  Accept/reject criteria for, 89-91
  Assurance, 28-29
  Compliance, 28
  Procedures, 27
  Purpose of, 23-29
  Similarity to special process
    audits, 81-82
Quality system requirements, 2-3

References, 153

Schedule for audit, 32
Scoring the audit, 48-50
Special process audits, 81-83
  Policy and protocol, 82
  Similarity to quality system
    audits, 81-82
  Worksheets, 82-83
Specifications, 5-21
  MIL-I-45208A — Inspection
    system requirements, 8-10,
    10-19, 20
  MIL-Q-9858A — Quality program
    requirements, 5-8, 10-19, 20

Supplier audit, 32-43
Supplier status report, sample of, 35

Terms used in auditing discipline,
  93-104

Unified supplier base audits, 88

Worksheets, audit, x, 45-57
  Addendum, 47-48
  Evaluation, 49, 50
  MIL-I-45208A/AQAP-4 audit
    worksheet, 131-151
  MIL-Q-9858A/AQAP-1 audit
    worksheet, 105-130
  Scoring, 48-50
  Validation of, 49